A Colour Atlas of

Clinical Genetics

Michael Baraitser

Consultant Clinical Geneticist
Hospital for Sick Children
Great Ormond Street
London

Robin M. Winter

Consultant Clinical Geneticist
Kennedy–Galton Centre, Harperbury Hospital
Radlett, Herts

and
Division of Inherited Metabolic Disease
Clinical Research Centre
Northwick Park Hospital
Harrow, Middx

Wolfe Medical Publications Ltd

General Editor, Wolfe Medical Atlases:
G. Barry Carruthers, MD(Lond)

Copyright © M. Baraitser and R.M. Winter, 1983
Published by Wolfe Medical Publications Ltd, 1983
Printed by Royal Smeets Offset b.v., Weert, Netherlands
ISBN 0 7234 1547 1
Paperback edition, © 1988

A CIP catalogue record for this book is available from the British Library.

For a full list of Wolfe Medical Atlases, plus forthcoming titles and details of
our surgical, dental and veterinary Atlases, please write to Wolfe Medical
Publications Ltd, 2-16 Torrington Place, London WC1E 7LT, England.

Contents

Acknowledgements

Dr Barbara Ansell
Dr Michael Bamford
Dr Nicholas Barnes
Professor Peter Beighton
Dr Caroline Berry
Dr Eric Blank
Dr Edward Brett
Dr John Burn
Professor Cedric Carter
Dr Nicholas Cavanagh
Dr Michael Connor
Dr Martin Crawfurd
Dr Clare Davison
Dr Nicholas Dennis
Dr Michael Dillon
Dr Dian Donnai
Mr Oliver Fenton
Professor Malcolm Ferguson-Smith
Mr John Fixsen
Dr Christine Garrett
Dr Francesco Gianelli
Dr David Grant
Dr Christine Hall
Dr Anita Harding
Professor Peter Harper
Professor John Harries
Dr Roger Hitchings

Dr Peter Husband
Dr Israel Kessel
Professor Michael Laurence
Dr James Leonard
Dr Barry Lewis
Dr Sheila Lewis
Dr Michael Lieberman
Dr Duncan Matthew
Dr Gerald McEnery
Professor Victor McKusick
Dr Patrick Mortimer
Dr Marcus Pembrey
Dr Michael Pope
Dr Michael Preece
Dr Michael Ridler
Dr Mary Rossiter
Dr Mary Seller
Dr David Siggers
Dr Joan Slack
Dr Rosemary Stephens
Mr David Taylor
Dr James Taylor
Dr Richard Watts
Mr Peter Webb
Dr John Wilson
Dr Mark Winter
Professor Otto Wolff

Without the generous help from the photographic departments at the Hospital for Sick Children, Great Ormond Street and Northwick Park Hospital, this Atlas could not have been produced.

Preface

Genetically determined conditions are individually rare but collectively they contribute significantly to the 2–3% of children born with malformations. If those conditions with a later age of onset are added to the total, it is not surprising that there are between 2000 and 3000 conditions listed by McKusick in his catalogue of *Mendelian Inheritance in Man* (see References). Unfortunately, the diagnosis of these conditions can be difficult. X-rays and blood tests are not always helpful and many of the syndromes can only be identified by pattern recognition. It is for this reason that pictures have become indispensable for the clinician and it is hoped that an Atlas will aid the diagnostic process by making typical representations available.

In the present work, verbal descriptions are brief and the reader is referred to other works for further details. Systems are dealt with individually but the section on syndromes defies precise grouping and the reader is encouraged to page through the section in order to search out the appropriate diagnostic category that his patient might have.

1 Introduction

Clinical genetics

Patients with genetic disorders can be encountered in every speciality of medicine; in their management, the clinical geneticist has a specific role which involves diagnosis, risk estimation and interpretation of laboratory results, and counselling.

a. The diagnostic role

Although a broad diagnosis may be made by a specialist, the clinical geneticist will appreciate the possibility of heterogeneity within a particular disease entity. It is important to recognize that the combination of signs and symptoms which make up a syndrome can have different causes. For example, motor neurone disease is usually sporadic (i.e. only one family member is affected), but in about 10% of cases there is a positive family history and inheritance may be autosomal dominant: retinitis pigmentosa can be caused by autosomal recessive, dominant or X-linked genes; diabetes is aetiologically heterogeneous – some types having a stronger genetic component than others. The different modes of inheritance might be distinguished by an assessment of factors such as age of onset, severity, and duration of the disease. If there are no specific pointers in an individual case, then the relative probability of a particular causative factor could be calculated.

The variability of genetic disorders within families must also be taken into account. Gene carriers may show minimal signs of a disorder, which nevertheless have important consequences for the risk to offspring, and extended family studies may be necessary in order to identify individuals at risk.

In certain groups of disorders the clinical geneticist has a primary diagnostic function. Syndromes caused by chromosomal abnormalities can often be diagnosed clinically; non-chromosomal malformation syndromes must be recognized when a family is sent for genetic counselling. Because of referrals for chromosomal analysis and genetic counselling, experience is gained of a great variety of syndromes, and this is reflected by the large number of 'new' syndromes reported in the specialized genetic literature.

b. Risk estimation and interpretation of laboratory results

The estimation of risks is fairly straightforward when one is dealing with a simple Mendelian situation, such as a mating between two known carriers of an autosomal recessive disorder or a mating when one partner is affected with an autosomal dominant disorder, however complications can arise in certain situations. For example, if a male is the only member of a family affected with an X-linked recessive disorder, then his mother might be a carrier of the abnormal gene; alternatively he could have received the abnormal gene as a new mutation, and in this case his mother would not be a carrier and would have a very low risk of having further affected sons. Estimation of precise risks that a woman is a carrier in this situation would depend upon the combination of various items of data, such as the number and relationship of normal males in the pedigree, and the results of any carrier detection tests (for example, the estimation of serum creatine phosphokinase in Duchenne muscular dystrophy). The methods of estimation are covered in many standard works on clinical genetics (see References). In other situations, the chances of an individual carrying an abnormal gene can be estimated by studying the joint inheritance of a *linked* marker gene in the family (see page 13). This approach has been used to estimate the chance of carrying the gene for dystrophia myotonica, both in individuals at risk and for fetuses prenatally, using linkage to the secretor locus for the ABO blood group substances.

Some risks cannot be based on Mendelian theory and must be obtained from large surveys of couples who have had a child with a particular abnormality. These *empirical* risks are used in counselling common malformations such as neural tube defects, dislocated hip or cleft lip and palate, where the causation is thought to be multifactorial.

c. Counselling

The essential prerequisites for good genetic counselling are an accurate diagnosis of affected family members (including an assessment of the specific mode of inheritance operating in the pedigree) and the correct estimation of the risks to other family members. These processes rely on the diagnostic and analytical skills of the clinical geneticist. Having decided on the specific risks, the geneticist must then communicate them to the patient, and this entails a different range of skills. It is important to determine the nature of the patient's enquiry. If parents have had an abnormal child they may wish to know the risks to future children. Alternatively they may come purely to discuss prognosis in the affected child or they may be concerned about their normal children and their future offspring. Occasionally considerable blame and guilt is expressed, either because individuals feel they are personally responsible for passing on an abnormal gene, or because parents feel that avoidable factors during a pregnancy gave rise to an abnormal child. These anxieties and guilt feelings should be discussed and as far as possible relieved.

Risks should be presented and put in perspective. There are many different ways of expressing risk. In racing circles odds are often used. These are the ratio of the probabilities of two distinct events. For example the odds for an abnormal child as against a normal child might be 1:3 (1 to 3). Risks for a particular event can be derived from odds. For example odds of an event A as against B of 1:3 represent a risk of A of 1 in $(1 + 3) = 4$. Risks can also be represented in percentage terms (e.g. 1 in 4 = 25%) or as absolute probabilities (i.e. 1 in 4 equals a probability of 0.25). In the authors' experience risks are best expressed as proportions (e.g. 1 in 2, 1 in 50, etc.), but it is important that patients get the risks the right way round. It is always worth emphasizing that a 1 in 20 risk of an abnormal child means that 19 times out of 20 one would expect a child *not* to be affected. The genetic counsellor rarely tells a patient directly what to do as a consequence of specific risks, although he does explain all the options. Nevertheless risks can be put in perspective and looked at in the light of the severity of the genetic condition. Everyone takes a 1 in 50 risk of having a child with a serious problem each time they reproduce. In the light of this, figure genetic risks of 1 in 50 are usually acceptable. In general terms, risks greater than 1 in 10 are high and less than 1 in 20 are considered low. Between 1 in 10 and 1 in 20 is an intermediate zone and risks in this area must be discussed in detail, in relation to the severity of the disease.

If parents want further children and risks are high, then several different strategies must be discussed. Depending on the circumstances these might include prenatal diagnosis (where available for the disorder in question), artificial insemination (donor) or adoption.

It is not the place of an Atlas of this nature to cover in detail the intricacies of risk estimation and counselling. A brief résumé of some basic genetic facts is given, but the bulk of the book is devoted to the clinical manifestations of genetic syndromes with particular emphasis on heterogeneity and syndrome identification.

2 Pedigree symbols

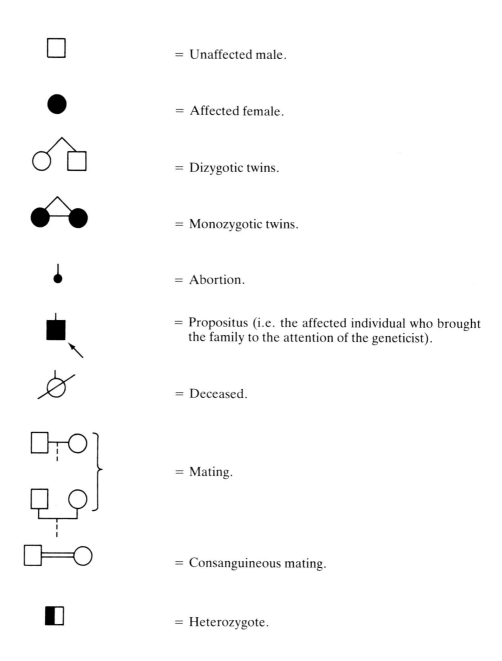

□ = Unaffected male.

● = Affected female.

= Dizygotic twins.

= Monozygotic twins.

= Abortion.

= Propositus (i.e. the affected individual who brought the family to the attention of the geneticist).

= Deceased.

= Mating.

= Consanguineous mating.

= Heterozygote.

3 Mendelian inheritance

Genes and chromosomes

1 Genes are carried on chromosomes. Because chromosomes are paired, everybody carries two copies of each gene (with the exception of genes carried on the sex chromosomes in males). Each gene is located at a specific point on a chromosome known as a locus.

The diagram shows a pair of chromosomes (*'homologues'*) on which are situated two loci. At each locus there are two genes, one on each chromosome.

2 Genes at the same locus are known as *alleles*. If the two alleles at a locus are identical, the individual is said to be *homozygous* at that locus. If the two alleles are non-identical, then the individual is said to be *heterozygous*.

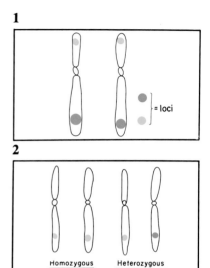

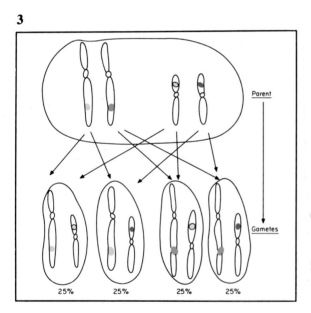

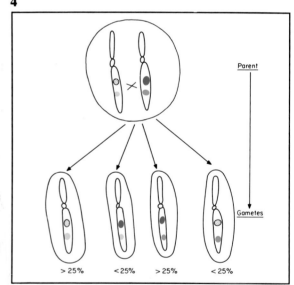

3 At meiosis the chromosome number is halved, with each member of a chromosome pair passing to a different gamete. The chance of an offspring inheriting a specific allele is therefore 1 in 2 (50%). This process is known as *segregation*. If genes are carried at loci on different chromosomes, then the inheritance of an allele at one locus is independent of the inheritance of an allele at the other locus. This is known as independent assortment.

4 When genes are carried on the same chromosome, there can only be independent assortment if a cross-over occurs between the loci at meiosis. The closer the loci are together the more unlikely this will occur. Two loci that are close together on a chromosome are said to be *linked*. Alleles at linked loci do not segregate independently. This means that pairs of alleles ('haplotypes') at linked loci on the same chromosome in a parent are more likely to be passed on together to offspring, rather than being split up by crossing-over.

5 **If two loci are linked**, it does *not* mean that alleles at one locus are more likely to be carried on the same chromosomes as alleles at the other locus, if one selects individuals at random from the population.

For example, the locus for the nail patella syndrome is linked to the ABO blood group locus (the nail patella syndrome is an autosomal dominant disorder). This does not mean that someone in the population with a particular blood group (e.g. AB) is more or less likely to have the nail patella syndrome. Nevertheless, in a particular family, the nail patella allele will tend to segregate with a particular ABO blood group allele.

The pedigree suggests that the nail patella allele is carried on the same chromosome as the blood group B allele in I_1; this can be inferred by looking at her children. Children of I_1 who inherit the B blood group allele also inherit the nail patella syndrome. II_2, however, has a child who is blood group O, but who has the nail patella syndrome. This means that a cross-over must have occurred between the two loci at meiosis during the formation of the ovum that gave rise to III_2.

If two characters do tend to be found together in members of the population, then this is known as *association*. Association of characters does not necessarily mean that they are caused by linked genes (although this can sometimes be the cause, if two loci are very closely linked). An example of association is the increased tendency of individuals with blood group O to develop duodenal ulcers. At present this phenomenon is not thought to be caused by a gene linked to the ABO blood group locus.

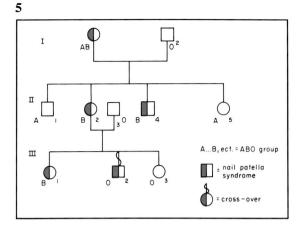

4 Types of pedigree

Autosomal dominant with complete penetrance

6

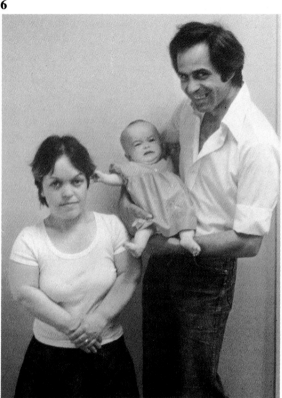

6 Achondroplasia in mother and daughter with normal father. This illustrates autosomal dominant inheritance.

7

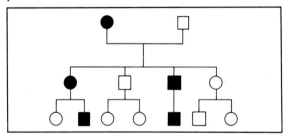

7 Note:
(a) Both males and females are affected.
(b) Affected individuals can pass the condition on to children of either sex; on average 1 in 2 children are affected.
(c) If an individual is not affected, then his children will show no signs of the condition.
(d) Male to male transmission.

8

Autosomal dominant inheritance with incomplete penetrance

8 Note that the segregation pattern is the same as complete penetrance, with the exception that some individuals show no signs of the disease, although they must carry the abnormal gene (II_1 and II_3).

Autosomal recessive inheritance

9

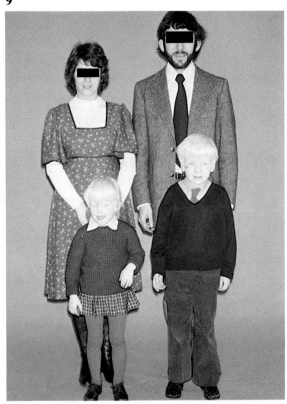

9 Oculo-cutaneous albinism in brother and sister. The parents, who are both heterozygous for the abnormal gene have normal pigmentation. This illustrates autosomal recessive inheritance.

Example: In certain populations, cystic fibrosis has an incidence of 1 in 1600 (= I) the gene frequency of the cystic fibrosis allele is therefore $\sqrt{I}$ = 1 in 40 and the carrier frequency is 2 × $\sqrt{I}$ = 1 in 20.

10

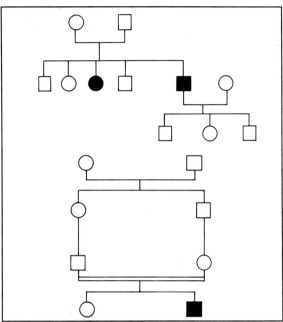

11

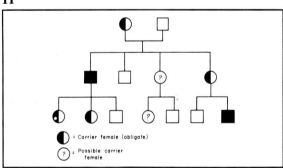

⬤ = Carrier female (obligate)

Ⓘ = Possible carrier female

X-linked recessive disorders

10 Note:
(a) Parents normal but sibs affected.
(b) Risk to further sibs 1 in 4.
(c) Consanguinity in parents is more common than in the general population. The rarer the disease the more likely parents are to be consanguineous.
(d) The risk to offspring of affected persons and their sibs is low. This is because the chance of someone marrying a carrier of the gene is small. If a disease is rare, then the gene frequency of the abnormal allele in the population is given by $\sqrt{I}$, where I is the incidence of the disease. The frequency of carriers of the abnormal gene is twice the gene frequency (because everyone carries two genes).

11 Note:
(a) Risk to male offspring of carrier females is 1 in 2.
(b) Risk of daughters of carrier females being carriers themselves is 1 in 2.
(c) If males reproduce *all* daughters will be carriers, no sons will be affected.
(d) There is no male to male transmission.
(e) An obligate carrier female is a woman with at least one affected son *and* at least one other affected male relative in the female line, or a female with two affected sons. Daughters of affected males are all obligate carriers.

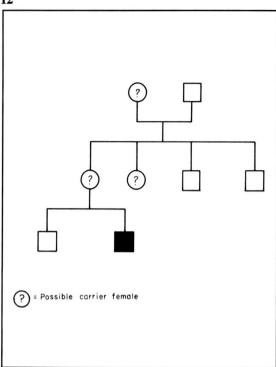

? = Possible carrier female

12 In isolated cases of known X-linked disorder some boys are the sons of carrier mothers and some are fresh mutants. When available, carrier detection tests on the mother could help solve the problem.

Consanguinity

If two individuals have at least one ancestor in common, they are said to be consanguineous.

If two consanguineous individuals marry, there is a chance that they will both carry identical genes passed down the pedigree from the common ancestor. The process is illustrated in **13**.

13

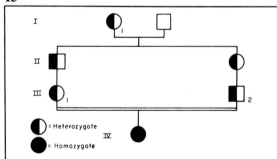

◑ = Heterozygote

● = Homozygote

I_1 is heterozygous for a recessive gene. Both III_1 and III_2 have received the gene from I_1; I_2 is also a common ancestor.

If III_1 and III_2 now have children, there is a possibility that the children will be homozygous for the recessive gene.

Genetic counselling

Studies have shown that on average, the number of harmful genes carried by each person is between one and two. For first cousins a risk figure of about 3–4% for a child with a recessive condition is usually given. The calculation of this risk is demonstrated in **14**.

14

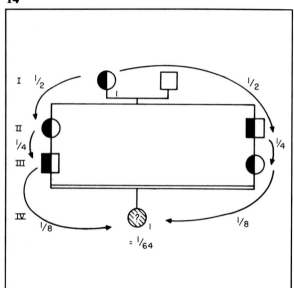

If I_1 carries a harmful recessive gene then the chance of IV_1 being homozygous for that gene, because of receiving a double dose of the gene from I_1 is 1 in 64. If everyone carries one harmful recessive gene, then IV_1 has a 1 in 32 chance of having an autosomal recessive syndrome (because IV_1 has two common ancestors, I_1 and I_2). If everyone carries two recessive genes then the risk is 1 in 16.

15

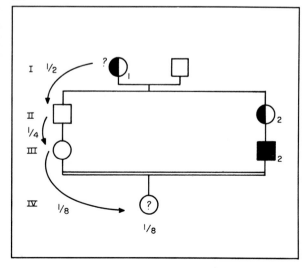

Where parents are consanguineous, and there is a member of the family affected by an autosomal recessive disorder, risk must be calculated from first principles (**15**). III_2 suffers from an autosomal recessive disorder. His mother II_2 must carry the gene, and so must one of his grandparents I_1 or I_2. From the grandparents there is a 1 in 4 chance that III_2's wife will receive the gene, giving a 1 in 8 chance that the offspring will be affected. Another example is shown in **16**.

16

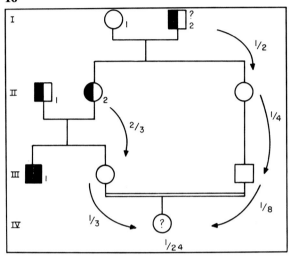

Multifactorial inheritance

Multifactorial (or polygenic) inheritance is used to denote the combined effects of many genes and the environment, or many genes alone (**17**). If an individual suffers from a disorder that is thought to be inherited in a multifactorial manner, then the following rules apply:

(a) The more severe the malformation, the greater the risk to sibs and offspring.

(b) If one sex is affected more frequently than another, then the risk to relatives of the less frequently affected sex is higher.

(c) If more than one individual in a family is affected, then recurrence risks are higher.

(d) The risk falls off rapidly as one passes from 1st to 2nd degree relatives.

Some of these rules can be illustrated using cleft lip and palate.

17

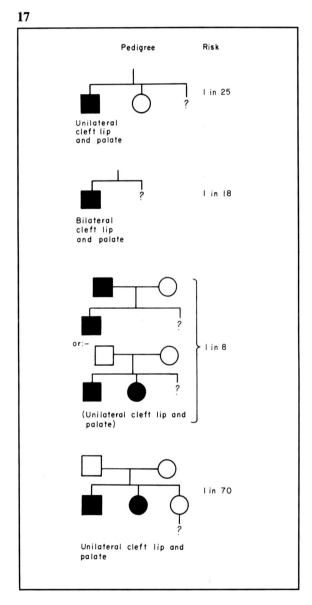

5 Chromosome nomenclature

Normal karyotype

In order to examine the chromosomes one must obtain dividing cells in culture (e.g. lymphocytes, fibroblasts, amniotic fluid cells). Cell division is usually arrested in metaphase and special staining techniques are used to visualize the chromosomes. The commonest technique is G-banding (Giemsa stain) which produces a pattern of positive and negative staining bands on the chromosomes.

18

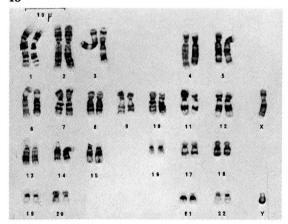

18 Normal male chromosomes. Individual chromosomes can be identified by their size and banding patterns. The chromosomes have been photographed, cut out and classified (a karyotype). There are 23 pairs (i.e. 46 chromosomes). The first 22 pairs are known as the autosomes and are numbered 1–22. The remaining pair consists of the sex chromosomes. In this case there is an X and a Y (a male). Normal females have two identical X chromosomes. The special staining technique has produced positive and negative bands on the chromosomes (G-banding).

Chromosome nomenclature

A chromosome is divided into long and short arms by a specialized area of the chromosome known as the *centromere*.

At metaphase; the chromosome has replicated and is ready for division; the centromere holds together the two identical parts of the chromosome (*the chromatids*). Band patterns are generally consistent for each chromosome and can be classified as shown in **19** and **20**.

19

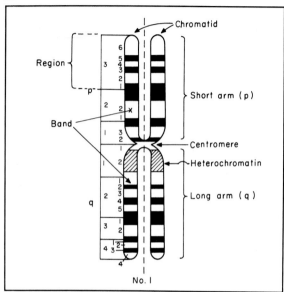

19 Chromosome nomenclature. Schematic drawing of a number 1 chromosome at metaphase. Regions and bands are classified on the left. A region is demarcated by either the centromere, the end of the chromosome or by a particularly prominent or constant band ('a land-mark band').

20

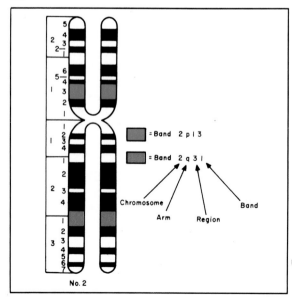

20 Band nomenclature.

Regions and bands are numbered from the centromere distally. A land-mark band is classified as belonging to the region that it demarcates distally.

Abnormalities of chromosome number

Each member of a chromosome pair comes either from the mother or the father. In order to achieve this the sperm and the egg must contain only one copy of each chromosome (i.e. they must contain only 23 chromosomes). The halving of chromosome numbers takes place at meiosis (reduction division) in the testis or ovary. After fertilization further cell division takes place by mitosis, where each daughter cell receives 46 chromosomes and is identical to the parent cell. Occasionally a daughter cell can receive an abnormal number of chromosomes. If this occurs at meiosis then an offspring with too many, or too few chromosomes can result.

[Examples of this are Down syndrome (trisomy 21) and Turner syndrome (monosomy X).]

21

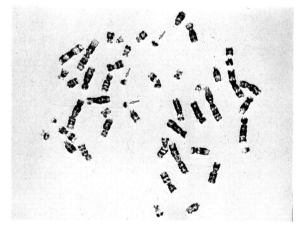

21 Down syndrome karyotype. In Down syndrome there are 47 chromosomes instead of 46. The extra chromosome is a number 21, giving three copies (trisomy 21). The figure shows a karyotype as seen down the microscope. The three chromosomes number 21 are arrowed.

Chromosome re-arrangements

In addition to abnormalities of chromosome number, individual chromosomes can be broken and re-arranged in such a way that their structure is altered. Several different types of re-arrangement are possible, e.g.

(a) Paracentric inversions (**22**).
(b) Pericentric inversions (**23**).
(c) Reciprocal translocations (**24** and **25**).
(d) Deletions (**26**).
(e) Isochromosomes (**27**).
(f) Ring chromosomes.

22

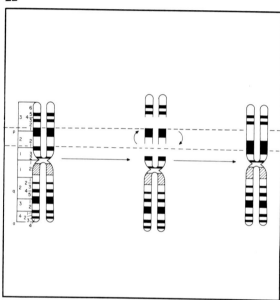

22 Paracentric inversion of chromosome 1 (i.e. an inversion of chromosome material not involving the centromere).

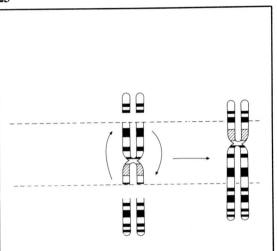

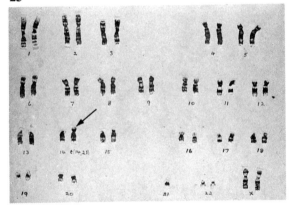

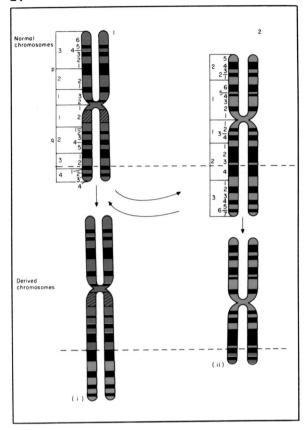

23 Pericentric inversion (i.e. the inversion involves the centromere).

N.B. Individuals carrying these chromosomes would have no pathological features, but there might be a risk to offspring.

24 Reciprocal translocation. Exchange of material has taken place between the long arms of chromosomes 1 and 2 leading to two derived chromosomes. As no material has been lost, this is a 'balanced' translocation. If an offspring inherited chromosome (i) without chromosome (ii) the infant would receive extra chromosome 2 material and a deficiency of chromosome 1 material (i.e. it would be partially trisomic for chromosome No. 2 and partially monosomic for chromosome 1).

Where a reciprocal translocation involves the centromeric region of chromosomes 13–15 or 21–22 it is known as a Robertsonian translocation.

25 Robertsonian translocation involving chromosomes 14 and 21 (arrowed). Note that although there are now only 45 chromosomes, no significant chromosome material has been lost – a 'balanced' translocation.

26

27

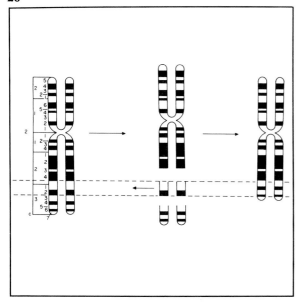

26 Chromosome deletion. Two breaks have taken place in the long arm of chromosome 2. The intervening material has been lost. This is known as an interstitial deletion. Individuals with this re-arrangement would most likely be phenotypically abnormal, as they would have an unbalanced amount of chromosome material.

27 Isochromosome. Normally a chromosome divides down the long axis during replication. If division occurs transversely through the centro-mere this results in the duplication of either the short or the long arm (an isochromosome). For illustrative purposes the figure shows the formation of an isochromosome for the short arms of chromosome 2. Most isochromosomes involve 13–15, 21–22 or the X.

Karyotype nomenclature

By convention, the number of chromosomes is given first, followed by the types of sex chromosomes, followed by the types of any additional, missing or abnormal chromosomes, e.g.

46, XX	– Normal female.
45, X	– Turner syndrome.
47, XY, + 21	– Male with trisomy 21.
69, XXY	– Triploidy, XXY sex chromosome compliment.
45, X/46, XX	– Mosaic Turner syndrome.

For inversions or translocations, the numbers of the chromosomes involved are given in brackets (with the chromosome with the lowest number first). This is followed by the bands or regions involved in further brackets. Symbols are used to indicate the type of re-arrangement involved.

28

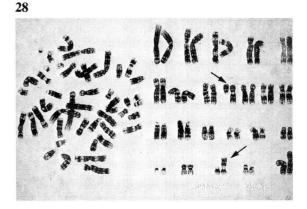

28 46, XX, t (9; 21) (q11; p11) karyotype. This means that chromosome 9 has broken at band q11 and chromosome 21 has broken at band p11. The small letter, t (for translocation), indicates that material has been exchanged between the two chromosomes (arrows).

6 Chromosome disorders

30

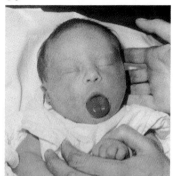

29

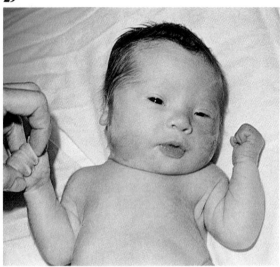

31

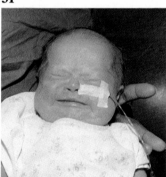

29–35   Down syndrome.
Note: Brachycephaly with flat occiput, up-slanting palpebral fissures, epicanthic folds, Brushfield spots, small nose with low nasal bridge and small ears with overfolded helix. Open mouth with protruding tongue. Single palmer crease with fifth finger clinodactyly; sandle gap between first and second toes. Increased risk of occurrence with advanced maternal age (e.g. 1 in 100 at 40 years; 1 in 50 at 45 years). Mothers over 37 years should be offered an amniocentesis.

32

Risk figures in Down syndrome	
	RISK
After one trisomic child (mother aged under 37 years)	1%
If mother aged over 37 years	Double maternal age risk
Mother is a 14/21 translocation carrier	10%
Father is a 14/21 translocation carrier	2%
One parent a 21/21 translocation carrier	100%

33

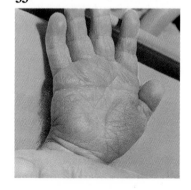

34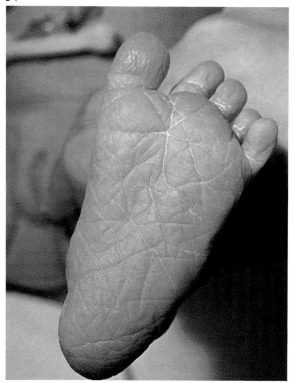

35

36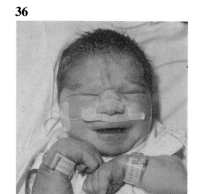

37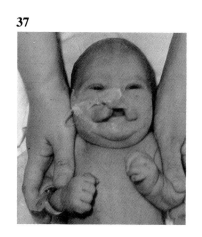

38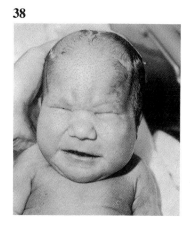

36–40   Trisomy 13.

Note: Microcephaly, low hairline, scalp defects, sloping forehead, microphthalmos, bilateral cleft lip and palate, broad nasal bridge, abnormal low-set ears and polydactyly with over-lapping fingers.

Other features: Holoprosencephaly, retinal dysplasia, capillary haemangiomata on face, cardiac anomalies, renal anomalies, hypospadias and cryptorchidism.

Severe mental retardation in survivors.

39

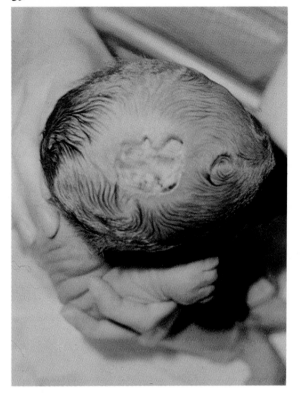

40

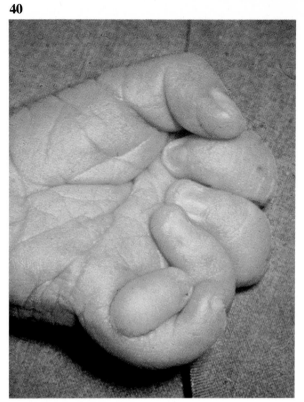

41

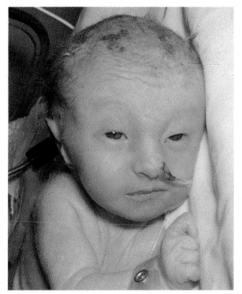

42

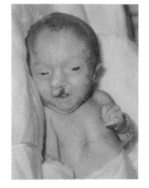

43

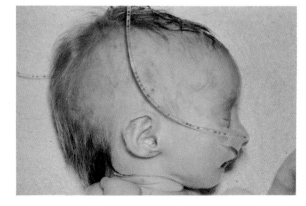

41–44 Trisomy 18.

Note: Small chin, low-set ears, prominent occiput, short palpebral fissures, and cleft lip and palate in some.

Wide-spread nipples, short sternum and narrow pelvis.

Overlapping fingers with small nails.

Other features: Cardiac anomalies, renal anomalies, cryptorchidism and severe mental deficiency.

44

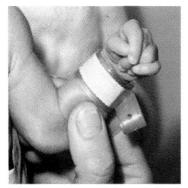

46

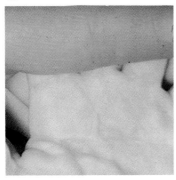

45

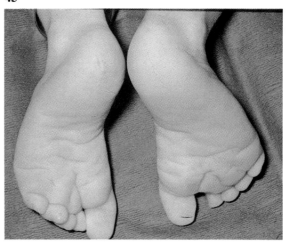

45 and 46   Partial trisomy 8.
Note: Deep palmar and plantar creases causing furrowing of the skin.

Other features: Long, narrow face, prominent, thick lower lip, skeletal malformations, absent patellae and moderate mental retardation.

Inheritance: Most cases are mosaic for trisomy 8/ normal cell lines.

47

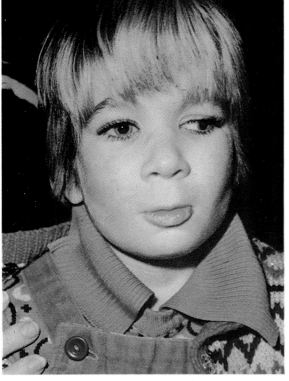

48

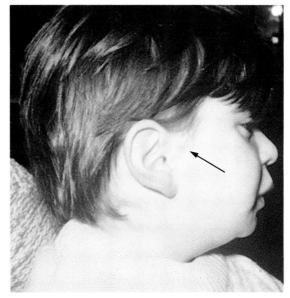

47 and 48   Trisomy 22.
Note: Long philtrum.

Pre-auricular pits, large ear and micrognathia.

Other features: Anal atresia and colobomata occur in partial trisomy 22 ('cat eye' syndrome).

49

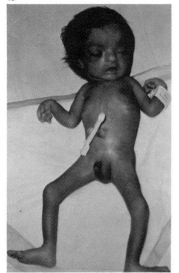

50

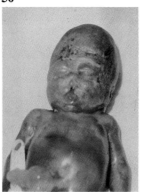

51

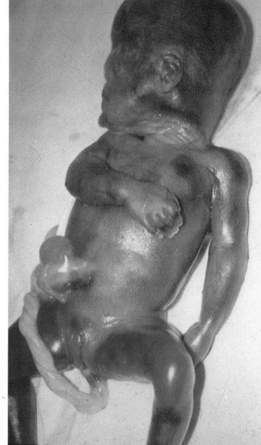

49–51   Triploidy.

Note: Short sternum, low-set ears, micrognathia, syndactyly of 3rd and 4th fingers and genital anomalies.

Hypotelorism, mid-line cleft lip and rudimentary nose (pre-maxillary agenesis).

Other features: Ocular colobomata, toxaemia during pregnancy and hydatidiform changes in the placenta. Most cases die early *in utero*.

Inheritance: Recurrence in a sibship is very rare, although there may be an increased risk for other chromosome aneuploidies.

52–54   9p Trisomy.

Note: Deep-set eyes, bulbous nose and thin upper lip.

Severe clinodactyly and brachymeso-phalangia of fifth fingers.

Other features: Mental retardation. Individuals with this condition have three copies of the short arm of chromosome 9.

52

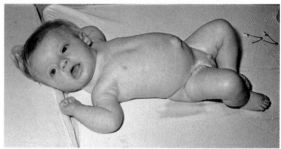

53

54

55

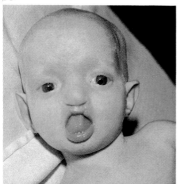

55 4p-Syndrome (Wolf–Hirschhorn syndrome).
Note: Hypertelorism and prominent glabella giving 'Greek warrior helmet' appearance, broad nasal tip, bilateral cleft lip, prominent lower lip, simple ears and iris colobomata.

Other features: Cleft palate, pre-auricular tags and pits, mental retardation and microcephaly. This syndrome is caused by a deletion in the short arm of chromosome 4.

56

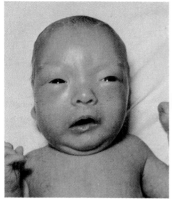

57

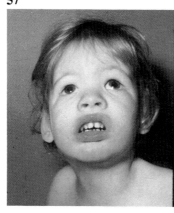

58

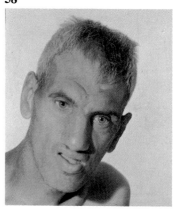

56–58 Cri du chat (5p-syndrome).
Note: Round face in the newborn, hypertelorism, epicanthic folds and strabismus.

Note elongation of the face in an older child and adult.

Other features: Low birth weight, 'cat-like' cry, mental retardation and low-set poorly formed ears.

59

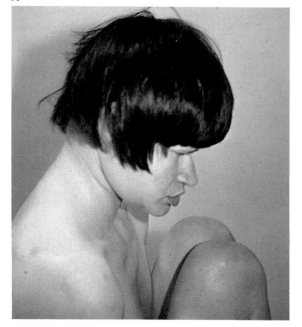

59 18q-Syndrome.
Note: Mid-face hypoplasia, prominent chin and everted lower lip.

Other features: Mental retardation, mild microcephaly, folded helices with prominent antitragus and atresia of external auditory canal. Long tapering fingers, dimples over extensor aspects of joints, abnormal genitalia and congenital heart defects.

60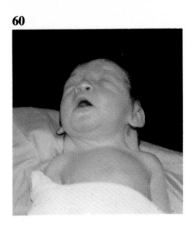

61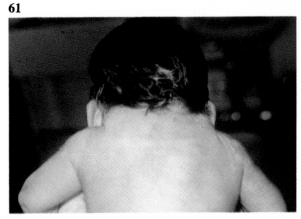

62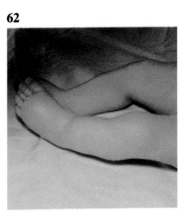

63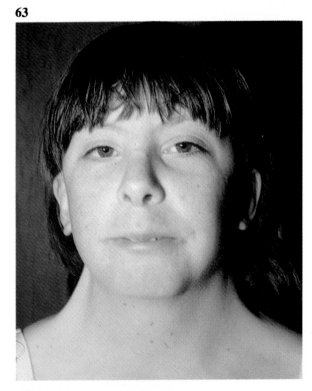

60–63 Turner syndrome.

Note: Webbing of neck with low hair-line ('trident'), low-set ears.

Hypoplastic nails and oedema of the dorsum of the hands and feet.

Other features: Short stature, shield like chest, cubitus valgus, primary amenorrhoea with streak gonads, pigmented naevi, co-arctation of the aorta and keloid formation. 45, XO chromosome constitution.

64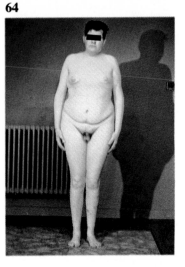

64 Klinefelter syndrome.

Note: 'Eunuchoid' habitus, increased carrying angle at elbow, gynaecomastia, small genitalia and female distribution of secondary sexual hair.

Other features: Mental retardation in only a few cases, infertility. 47, XXY chromosome constitution.

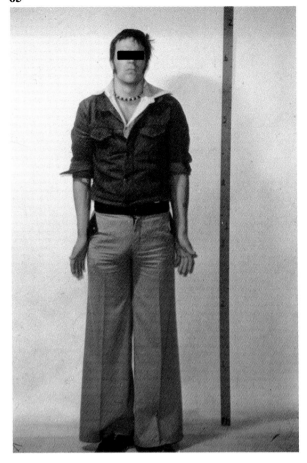

65 XYY syndrome.
Note: Tall stature, muscular build.
  Other features: Impulsive behaviour. Although there is a higher proportion of individuals with this condition in correctional institutions than in the general population, the exact incidence of psychopathic behaviour is probably low. 47, XYY chromosome constitution.

66 XXX syndrome.
Note: Slender body habitus.
  Other features: Average IQ is about 30 points lower than in 46, XX females.

67

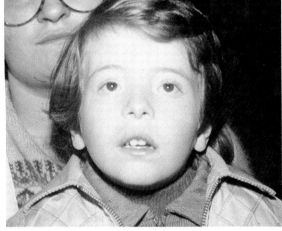

67 XXXXY syndrome.
Note: Wide-set eyes, low-set nasal bridge, epicanthus, mandibular prognathism, low-set ears and short neck.
  Other features: Mental and motor retardation, radio-ulnar synostosis, cryptorchidism and small penis and congenital heart defect.

68

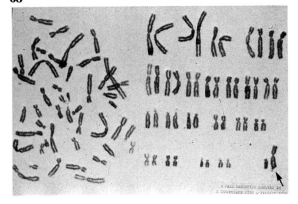

69

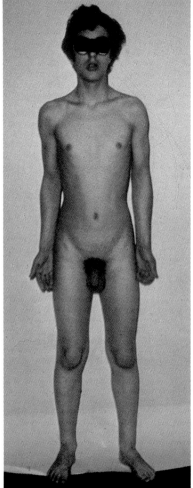

70

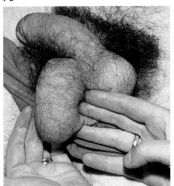

68–70 X-linked mental retardation with fragile sites (Martin–Bell syndrome).

Note: Thin face, prominent jaw and enlarged testes.

Fragile site at the end of the long arms of X-chromosome (arrow).

Other features: Mental retardation, speech delay. Female carriers are sometimes retarded.

Inheritance: X-linked recessive.

7 Dysmorphic syndromes

71

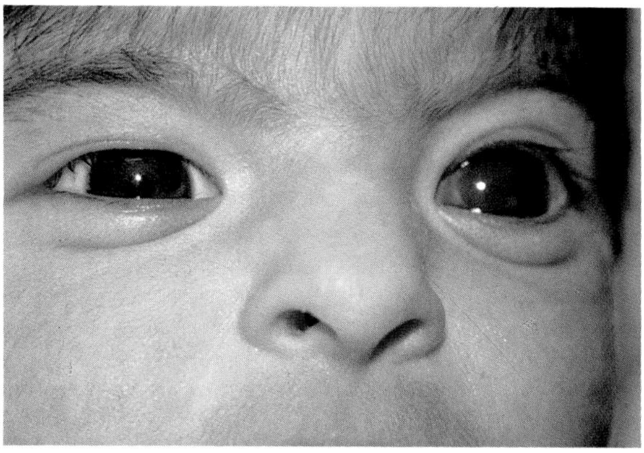

72

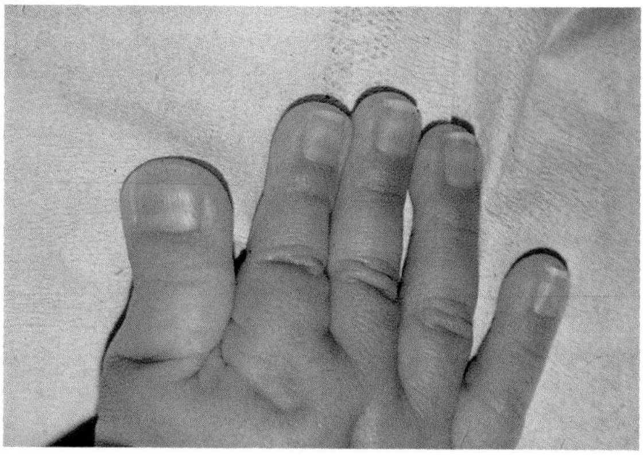

73

74

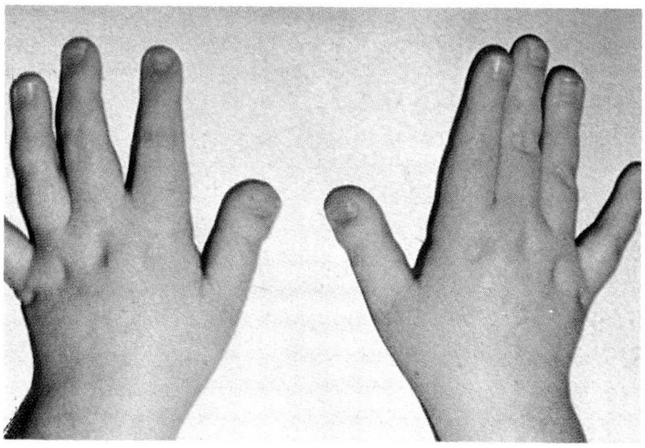

75

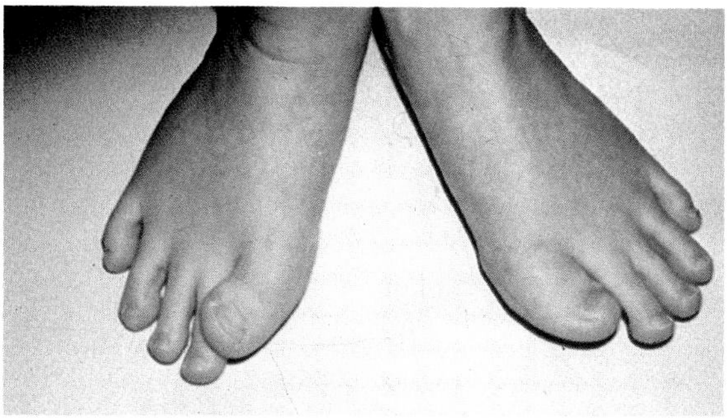

76

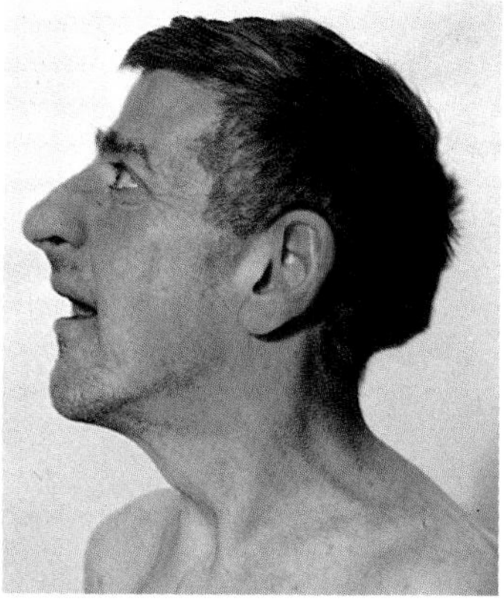

71–80   Rubinstein–Taybi syndrome.

Note: Mild microcephaly, antemongoloid slant to eyes with ptosis, glaucoma and strabismus, prominent hooked nose with broad nasal bridge and nasal septum extending below alae. Broad or duplicated terminal phalanges of thumb and halluces. Generalized hirsutism.

Other features: Mental retardation, short stature.

Inheritance: Most cases sporadic.

77

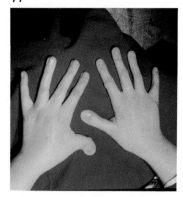

78

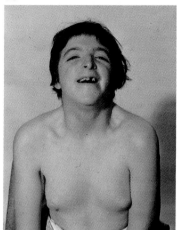

79

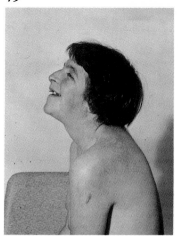

80

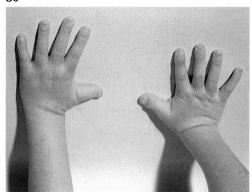

81

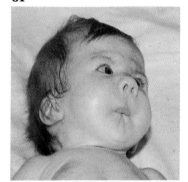

82

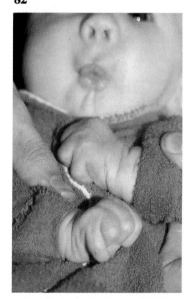

81 and 82 Freeman–Sheldon syndrome.
Note: Full forehead, blepharophimosis, hypoplastic alae, long philtrum, small mouth with puckered lips, H-shaped depression below lower lip. Ulnar deviated fingers with contractures ('Windmill vane'), talipes equinovarus with contractures of toes and vertical talus.

Other features: Kyphoscoliosis.

Inheritance: Mostly autosomal dominant. Occasional autosomal recessive pedigrees described.

83

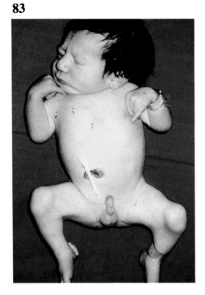

84

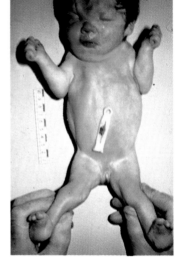

83 and 84   Multiple pterygium syndrome.
Note: Pterygia (webs) of knees and elbows, short neck, mid-face haemangioma, hypertelorism, epicanthic folds, flattened nose, talipes equinovarus, scoliosis, rocker bottom feet and deformed chest.

Other features: Cleft palate.
Inheritance: Autosomal recessive. There may be mild and severe forms.

85

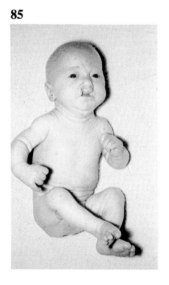

86

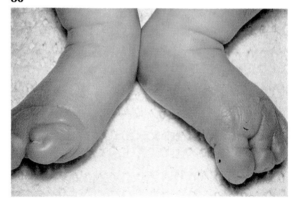

87

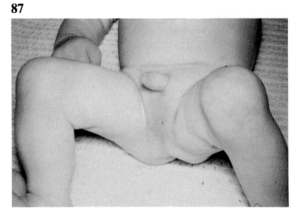

85–87   Popliteal web syndrome.
Note: Web of skin in popliteal region, hypoplastic scrotum, bilateral cleft lip/palate and dysplastic nails with V-shaped encroachment of the skin.

Other features: Ankyloblepharon, lower lip pits, and talipes equinovarus.

Inheritance: Mostly autosomal dominant pedigrees have been described although affected sibships with apparently normal parents have also occurred.

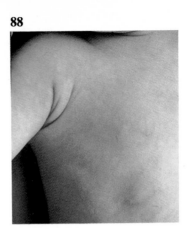

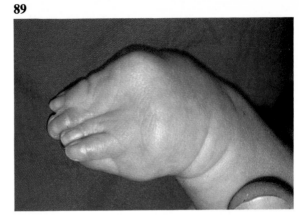

88 and 89   Poland syndrome.
Note: (a) Absence of sternal portion of pectoralis major on one side. (b) Syndactyly and hypoplasia of hands.

  Other features: Other hand and limb anomalies, rib anomalies.

  Inheritance: Usually sporadic.

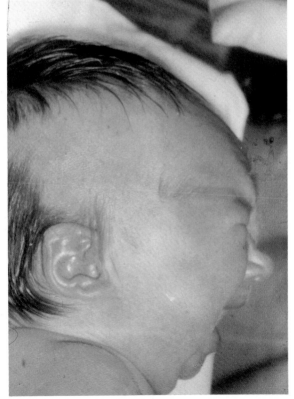

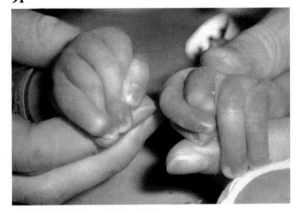

90 and 91   Beal syndrome (contractural arachnodactyly).
Note: Crumpled helix of the ear with a small notch. Long thin fingers with contractures.

  Other features: Contractures of other joints, talipes equinovarus and kyphoscoliosis. The contractures tend to improve with age.

  Inheritance: Autosomal dominant.

92

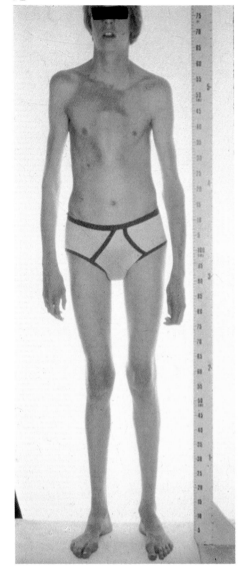

93

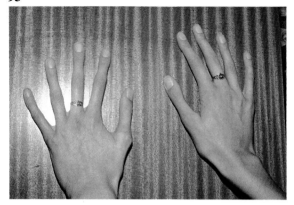

94

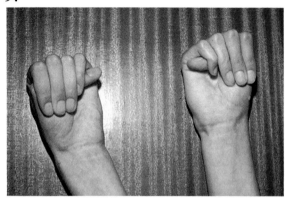

95

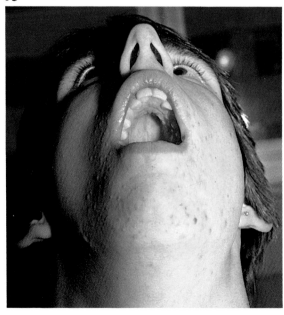

92–95   Marfan syndrome.
Note: Tall stature, reduced upper segment to lower segment ratio, pectus carinatum, long fingers and toes and high-arched palate.

Other features: Joint laxity, lens dislocation, dilatation of aortic root with regurgitation, floppy mitral valve and dissecting aneurysm.

Inheritance: Autosomal dominant.

96

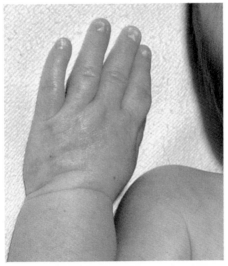

98

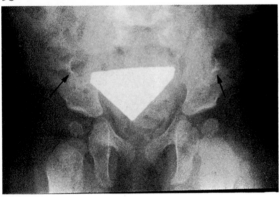

97

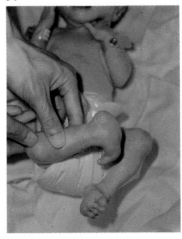

99

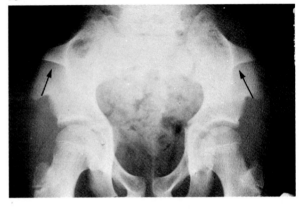

96–99   Nail patella syndrome.
Note: Joint contractures and club feet, dystrophic nails with triangular lunules and iliac spurs (arrowed) in a neonate and adult.
   Other features: Hypoplastic patellae, ptosis, cloverleaf pigmentation of the iris and nephropathy.
   Inheritance: Autosomal dominant.

100 and 101   Ectrodactyly, ectodermal dysplasia and clefting (EEC) syndrome.
Note: Repaired cleft lip, sparse, dry hair and split hand with partial syndactyly.
   Other features: Small or missing teeth.
   Inheritance: Autosomal dominant.

100

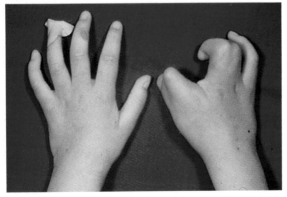

101

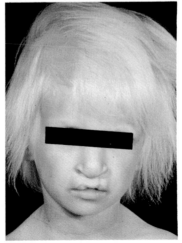

102

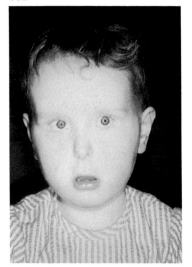

103

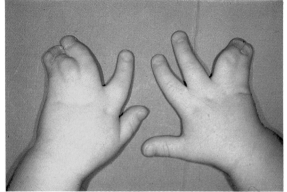

104

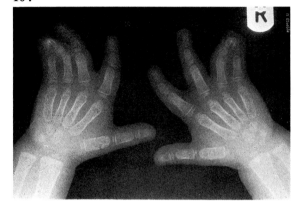

102–104 Oculo-dento-digital syndrome.
Note: Thin hypoplastic alae with small pinched nose, small eyes, irregular teeth and fine sparse hair.

Syndactyly of 3rd, 4th and 5th fingers with camptodactyly.

Other features: Enamel hypoplasia, normal intelligence.

Inheritance: Autosomal dominant.

105

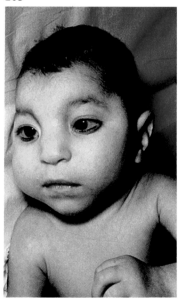

105 Cerebro-ocular-facio-skeletal syndrome (COFS).
Note: Microcephaly, prominent nasal bridge.

Other features: Joint contractures, microphthalmia, cataracts, severe progressive mental retardation and cachexia.

Inheritance: Autosomal recessive.

106

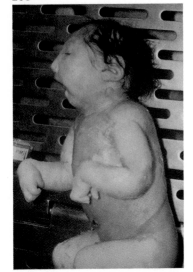

107

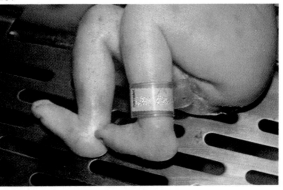

108

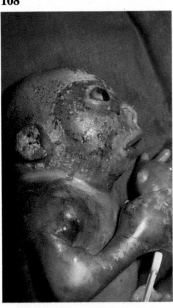

109

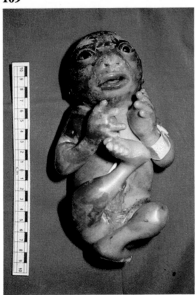

110

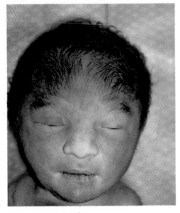

106–109   Neu-Laxova syndrome.
Note: Microcephaly, prominent nasal bridge, small chin, joint contractures, subcutaneous oedema, 'collodion' skin, rocker bottom feet and oedema of hands and feet.

Other features: Neonatal death, agenesis of the corpus callosum. Some infants with this condition have bizarre facial features with absent eyelids and nose. The hands and feet can be grossly swollen.

Inheritance: Autosomal recessive.

110 and 111   Pena Shokeir syndrome.
Note: Facial haemangioma, depressed nasal tip and hirsute forehead.

Gross joint contractures.

Other features: Polyhydramnios, pulmonary hypoplasia and neonatal death.

Inheritance: Autosomal recessive.

111

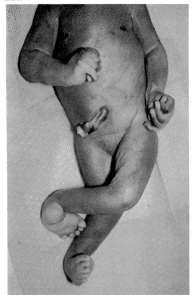

112

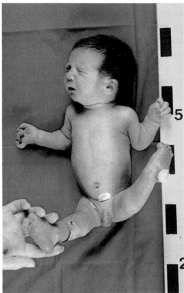

112 Larsen syndrome.

Note: Flat mid-face, depressed nasal tip, dislocation of knee and hip, positional deformities of the feet.

Other features: Prominent forehead, hypertelorism, cleft palate, short spatulate thumbs, short metacarpals and double ossification centres of the calcaneus.

Inheritance: Autosomal dominant and recessive forms with the latter being more severe.

113

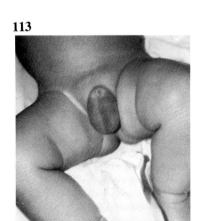

114

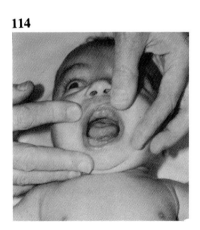

115

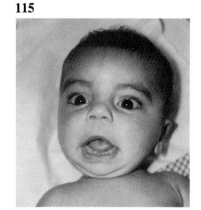

113–115 Robinow syndrome (fetal face syndrome).

Note: Broad high forehead, apparent hypertelorism, upturned nose with anteverted nostrils, long philtrum, large triangular-shaped mouth and gingival hypertrophy. Mesomelia of upper limbs, stubby hands and micropenis.

Other features: Hemivertebrae, intelligence usually normal.

Inheritance: Autosomal recessive and autosomal dominant families described.

116

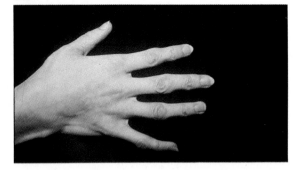

117

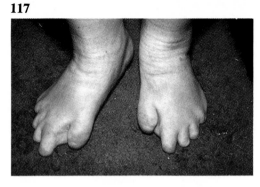

118

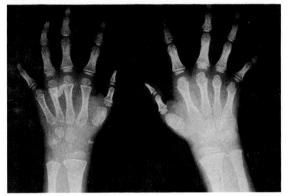

119

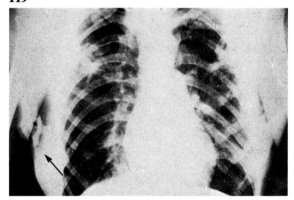

116–119   Fibrodysplasia ossificans progressiva.
Note: Short hallux with valgus deformity and abnormalities of the other toes. Short first metacarpal and hypoplasia of the phalanges. Note ossification in the soft tissues of the chest wall (arrowed).

Inheritance: Autosomal dominant pedigrees described, although most cases are sporadic.

120   Vater association.
Note: Bilateral radial aplasia.

Other features: Vertebral anomalies, anal atresia, tracheo-eosophageal fistula, radial and renal dysplasia. Congenital heart disease and non-radial limb defects may also be present.

Inheritance: Sporadic, 2% recurrence risk.

120

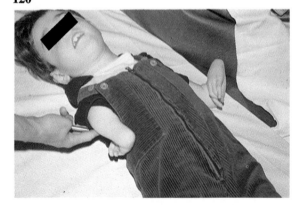

121

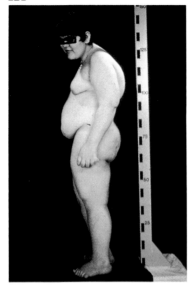

122

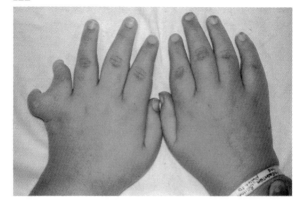

124

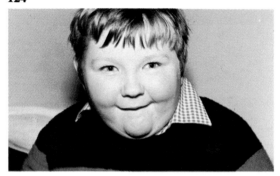

123

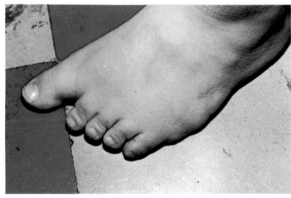

121–124 Laurence–Moon–Biedl syndrome.
Note: Obesity, polydactyly of hands and feet (extra digit removed from foot).

Other features: Mental retardation, retinitis pigmentosa, hypogonadism and renal defects.

Inheritance: Autosomal recessive.

125

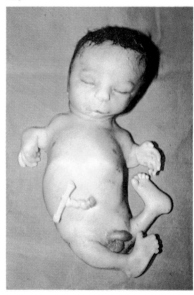

125 Roberts syndrome.
Note: Short deformed limbs, hypertelorism, depressed nasal tip, micrognathia, small cleft in upper lip, long philtrum and prominent premaxilla, limb defects, partial syndactyly of digits and enlarged genitalia.

Other features: Early death, mid-facial capillary haemangioma and malformed ears. Abnormalities of the chromosomes consisting of 'chromosome puffs'. This syndrome is probably identical to the 'pseudothalidomide syndrome'.

Inheritance: Autosomal recessive.

126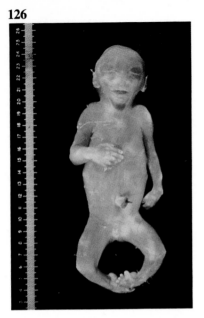

127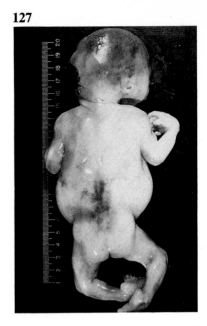

126 and 127   Meckel syndrome.
Note: Polydactyly, microcephaly, occipital encephalocele and talipes equinovarus.

Other features: Cystic kidneys, eye defects and cleft palate.
Inheritance: Autosomal recessive.

128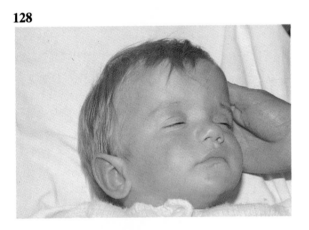

129

130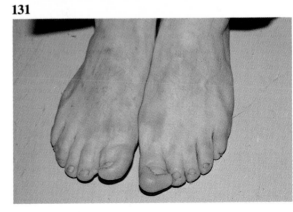

131

128–131   Greig cephalopolysyndactyly syndrome.
Note: Duplication of the big toe with syndactyly of toes 2–4. Post-axial polydactyly of hands, prominent forehead with a flat nasal bridge and mild hypertelorism.
Other features: Occasional mental retardation.
Inheritance: Autosomal dominant with very variable expression.

132 Terminal transverse limb defects.
Note: Transverse defects of all four limbs with no bony digital remnants.

Other features: Hypoplasia of the tongue should be excluded in order to rule out the Hanhart and hypoglossia-hypodactylia syndromes.

Inheritance: Usually sporadic.

132

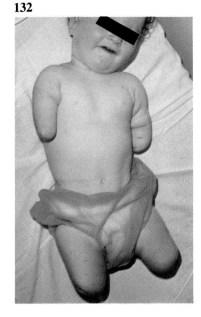

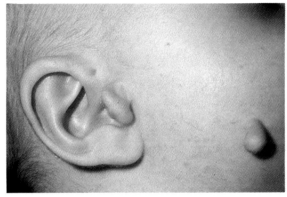

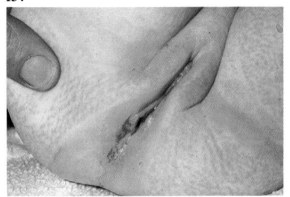

133 and 134 Townes syndrome.
Note: Pre-auricular skin tags, imperforate anus with a recto-vaginal fistula.

Other features: Triphalangeal or hypoplastic thumbs, lop ('satyr') ears.

Inheritance: Autosomal dominant with very variable expression.

135 Holt–Oram syndrome.
Note: Severe limb-reduction, hypoplastic thumbs. (Terminated fetus, child of an affected parent who had only mild limb defects.)

Other features: Triphalangeal thumbs, hypoplastic radius and other upper limb defects. Cardiovascular malformations, particularly ASD and VSD.

Inheritance: Autosomal dominant with very variable expression.

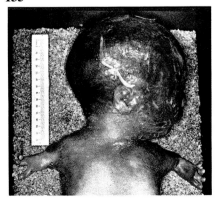

136

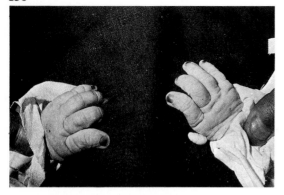

137

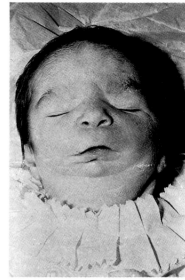

136 and 137   Nager syndrome.
Note: Malar and mandibular hypoplasia, deficient eyelashes, low-set ears and hypoplastic thumbs.

Other features: Cleft palate, deafness and upper limb anomalies.

Inheritance: Most cases are sporadic.

138

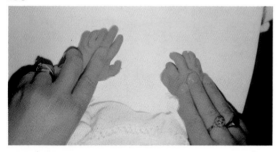

139

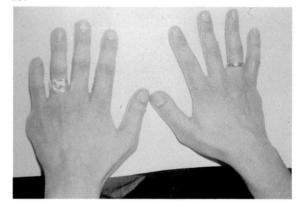

138 and 139   Synpolydactyly.
Note: Duplication of fourth fingers with partial syndactyly.

Similar features in the father, in whom the most lateral finger has been removed on both hands.

Other features: Synpolydactyly refers to the situation where syndactyly can occur in some individuals of a family without polydactyly, and not vice versa. Polysyndactyly refers to the opposite situation where polydactyly can occur without syndactyly.

Inheritance: Autosomal dominant.

140

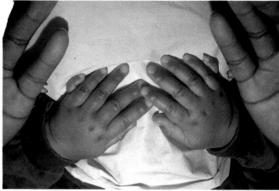

140   Post-axial polydactyly.
Note: Pedunculated extra digits in parent and child.

Other features: Nil.

Inheritance: Autosomal dominant.

141 Split hand (ectrodactyly).
Note: Missing digits from middle ray of hand with partially split palm giving a 'lobster claw' appearance.

Inheritance: Autosomal dominant with markedly reduced penetrance.

141

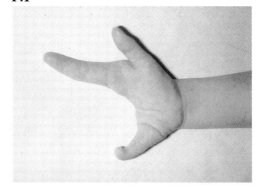

142

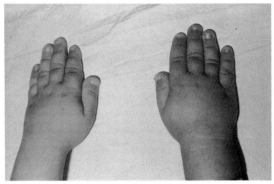

143

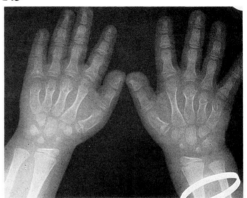

145

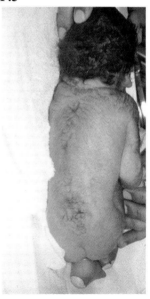

142 and 143 Brachydactyly type E.
Note: Short, stubby fingers, short nails, cone-shaped epiphyses of the phalanges and short metacarpals.

Other features: Short stature. Other syndromes such as pseudohypoparathyroidism and pseudopseudohypoparathyroidism should be excluded.

Inheritance: Autosomal dominant.

144

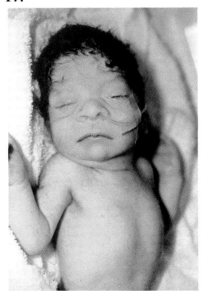

144–151 Cornelia deLange syndrome.
Note: Low birth weight, short stature, microcephaly, generalized hirsutism, synophrys, long eyelashes, anteverted nostrils, long philtrum, thin upper lip, micrognathia and downturned angles of mouth. Malformation of limbs in 20–30%.

Other features: Mental retardation, hoarse cry.

Inheritance: Usually sporadic. Empiric recurrence risk 2–3%.

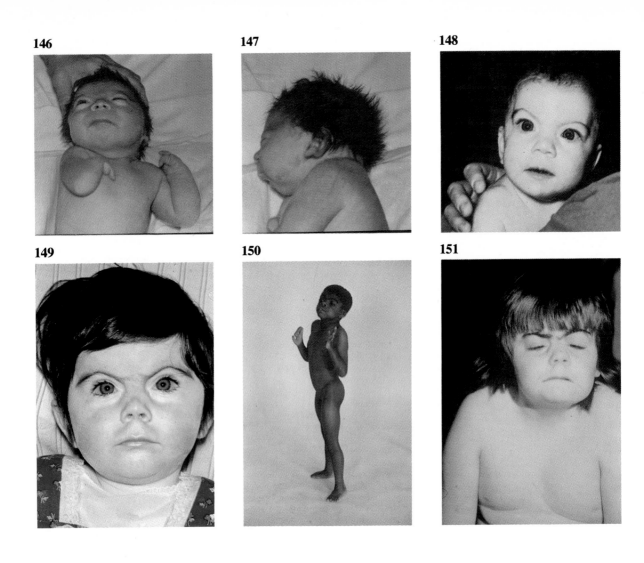

146   **147**   **148**

149   **150**   **151**

152

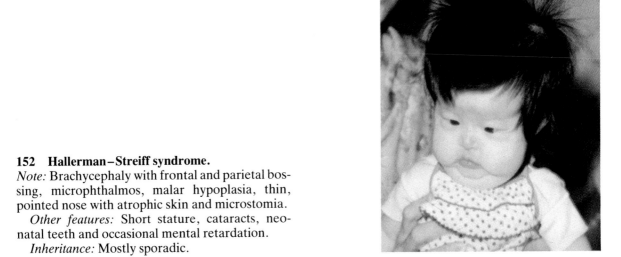

152   Hallerman–Streiff syndrome.
Note: Brachycephaly with frontal and parietal bossing, microphthalmos, malar hypoplasia, thin, pointed nose with atrophic skin and microstomia.

   Other features: Short stature, cataracts, neonatal teeth and occasional mental retardation.

   Inheritance: Mostly sporadic.

153

154

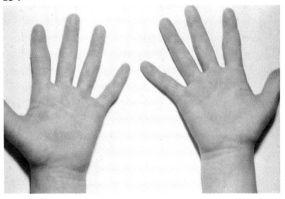

153 and 154   Tricho-rhino-phalangeal syndrome.
Note: Bulbous, pear-shaped nose, prominent philtrum, protruding ears and hypoplastic nares.

Abnormal angulation of phalanges with broadening of the middle phalangeal joint.

Other features: Mild growth deficiency, sparse thin hair. X-ray changes consisting of cone-shaped epiphyses in the hands.

Inheritance: Autosomal recessive and dominant forms.

155

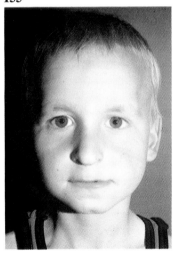

156

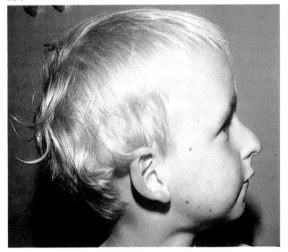

155 and 156   Langer–Giedion syndrome.
Note: Features similar to tricho-rhino-phalangeal syndrome.

Other features: Features which distinguish the two syndromes include microcephaly, mental retardation, multiple exostoses, loose and redundant skin at birth all of which occur in the Langer–Giedion syndrome.

Inheritance: Autosomal dominant.

157

159

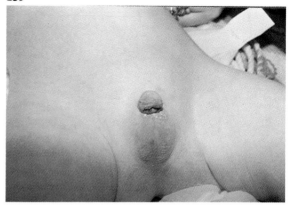

158

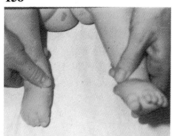

157–159 Smith–Lemli–Opitz syndrome.
Note: Microcephaly, bilateral ptosis, anteverted nostrils and longish philtrum (not shown well because of the angle of the photograph).
Syndactyly of 2nd and 3rd toes.
Hypoplastic male genitalia.
Inheritance: Autosomal recessive.

160

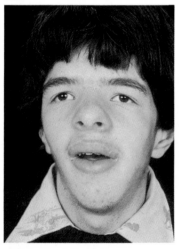

161

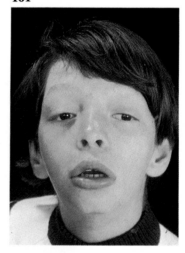

162

163

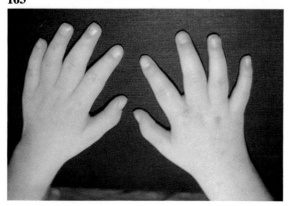

160–163 Coffin–Lowry syndrome.
Note: Downsloping palpebral fissures, bulbous nose with anteverted nostrils and thick alae, thick lips, prominent mid-line groove to tongue, long, soft and tapering fingers.
Other features: X-rays show tufting of the terminal phalanges and vertebral defects, mental retardation.
Inheritance: X-linked recessive with some expression in female carriers.

164

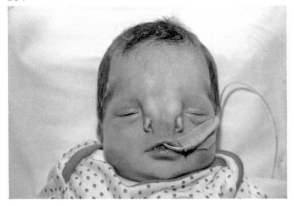

165

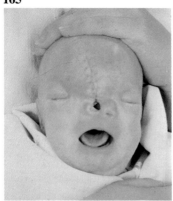

164 and 165   Fronto-nasal dysplasia.
Note: Hypertelorism associated with a bifid nasal tip or complete mid-line splitting of the nose.
  Other features: Median cleft palate, anterior encephalocele.
  Inheritance: Usually sporadic.

166

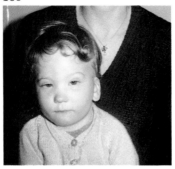

166   Blepharophimosis syndrome.
Note: Hypertelorism with marked epicanthus, ptosis.
  Inheritance: Autosomal dominant inheritance.

167

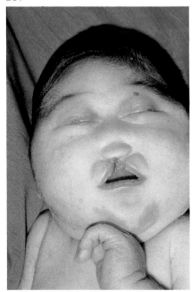

168

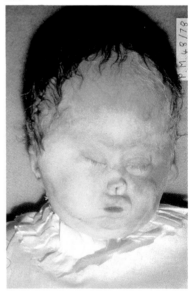

167–170   Holoprosencephaly.
Note: Range of facial abnormalities from premaxillary agenesis through cebocephaly to cyclopia and ethmocephaly.
  Bilateral or mid-line cleft lip extending into nose with hypotelorism (premaxillary agenesis).

Hypotelorism with single nostril (cebocephaly).
  Cyclopia with single central proboscis, lack of division of forebrain.
  Inheritance: Mostly sporadic. Chromosomal abnormalities should be excluded (e.g. Tri 13, 18p-, 4p-).

169

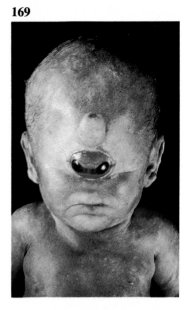

170

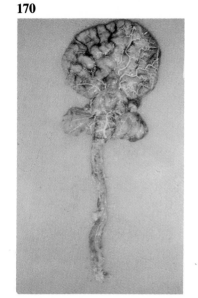

171

172

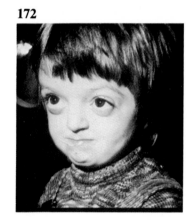

171 and 172   Crouzon syndrome.
Note: Prominent forehead, proptosis, hypertelorism, short upper lip, maxillary hypoplasia and hooked nose.

Other features: Brachycephaly, nystagmus and optic nerve damage and mental retardation only occasionally.

Inheritance: Autosomal dominant.

173

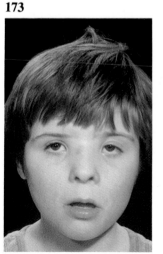

174

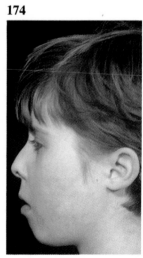

173 and 174   Saethre–Chotzen syndrome.
Note: Broad, flat nasal bridge, telecanthus, maxillary hypoplasia, prominent ear crus and facial asymmetry.

Other features: Craniosynostosis, occasional mental retardation and minor cutaneous syndactyly.

Inheritance: Autosomal dominant.

175

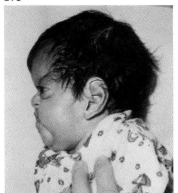

176

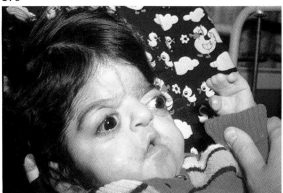

177

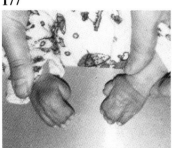

175–177 Apert syndrome.

Note: Brachycephaly, acrocephaly, proptosis, beaked nose, mid-facial hypoplasia, antemongoloid slant to eyes, syndactyly of all fingers giving 'base-ball glove' appearance, big toes and thumbs are often free.

Other features: Craniosynostosis, mental retardation, narrow or cleft palate and broad thumbs in valgus position.

Inheritance: Autosomal dominant, most cases are fresh mutations.

178

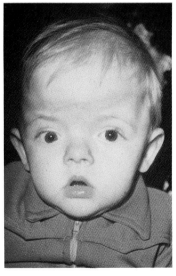

179

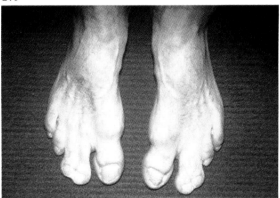

178 and 179 Pfeiffer syndrome.

Note: Acrocephaly, tall, prominent forehead, hypoplastic supra-orbital ridge, antemongoloid slant to eyes with hypertelorism, flat nasal bridge, broad great toes in valgus and syndactyly of toes 2–3 (in grandfather).

Other features: Craniosynostosis.

Inheritance: Autosomal dominant with very variable expression.

180   Hard ± E syndrome.
Note: Hydrocephalus, microphthalmia.
Other features: <u>H</u>ydrocephalus with lack of gyri (<u>a</u>gryia), <u>r</u>etinal <u>d</u>ysplasia with anterior chamber defects, <u>e</u>ncephalocele and retardation.
Inheritance: Autosomal recessive.

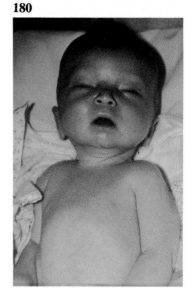

180

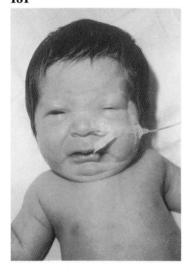

181

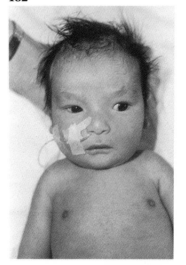

182

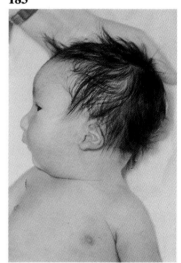

183

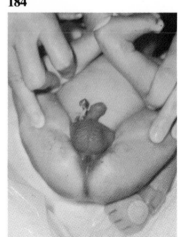

184

181–184   Opitz G syndrome.
Note: Hypertelorism, posteriorly rotated ears.
   Anteriorly placed anus.
   Other features: Hypospadias, swallowing difficulties and laryngeal clefts.
   Intelligence – usually normal.
   Inheritance: Probably autosomal dominant.

   N.B. This syndrome shows overlapping features with the Opitz BBB syndrome in which swallowing difficulties are not encountered.

185   Moebius syndrome.
Note: Bilateral facial weakness and bilateral abducens palsy.
Other features: Limb anomalies, including brachydactyly-syndactyly. Mental retardation, atrophy of tongue and swallowing difficulties.
Inheritance: Mostly sporadic, 2% recurrence risk.

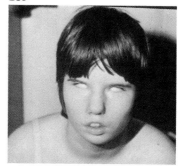

186   Johanson–Blizzard syndrome.
Note: Mild microcephaly, hypoplastic alae nasae.
Other features: Scalp defects, sparse hair, hydronephrosis, imperforate anus, hypothyroidism, pancreatic insufficiency and mild mental retardation.
Inheritance: Autosomal recessive.

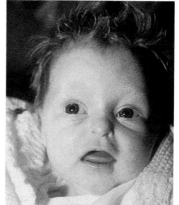

187   Schwartz–Jampel syndrome.
Note: Blepharophimosis, long eyelashes, expressionless face and difficulty in opening mouth.
Other features: Short stature, myopia, muscle wasting with myotonia and joint limitation.
Inheritance: Autosomal recessive.

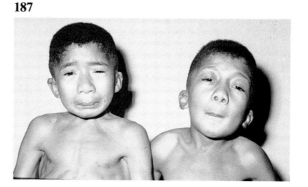

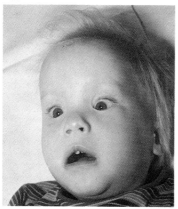

188 and 189   Plott syndrome.
Note: Prominent forehead, anteverted nostrils and mild facial diplegia.
Other features: Congenital laryngeal stridor due to abductor paralysis, mental retardation.
Inheritance: X-linked recessive.

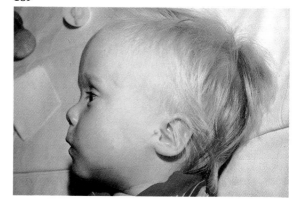

190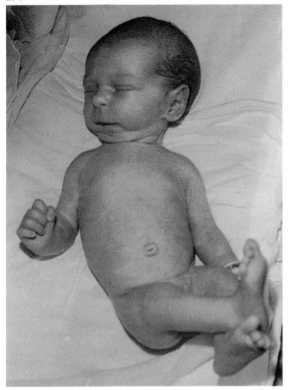

191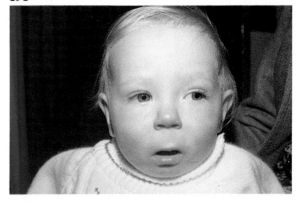

192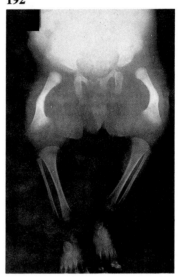

190–192 Femoral hypoplasia unusual facies syndrome.
Note: Short, stubby nose, anteverted nostrils, long philtrum, thin upper lip and micrognathia.
  Shortening and bowing of the legs.
  Shortening and angulation of the femurs.
  Other features: Mid-line cleft palate.
  Inheritance: Usually sporadic, but one dominant pedigree has been described. Maternal diabetes should be excluded.

193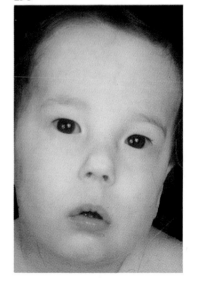

193–196 Marden–Walker syndrome.
Note: Short palpebral fissures, expressionless face, depressed nasal bridge, anteverted nostrils and low-set ears.
  Other features: Contractures of fingers and other joints, cleft palate, congenital heart disease and mental retardation.
  Inheritance: Autosomal recessive.

194

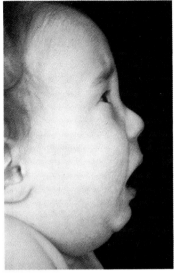

195

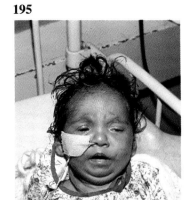

196

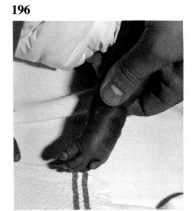

197

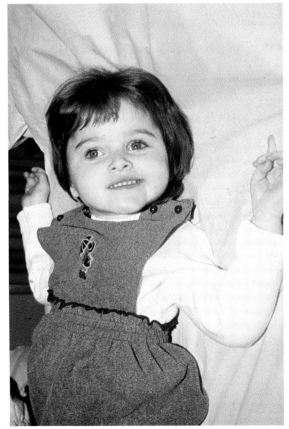

198

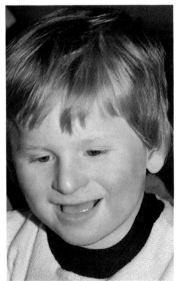

197 and 198   Happy puppet syndrome (Angelman).
Note: Deep-set eyes, prominent jaw, happy appearance and characteristic posture of arms.

Other features: Hypotonia, mental retardation, optic atrophy or choroidal dystrophy and seizures. Characteristic jerky movements are reminiscent of a string puppet.

Inheritance: Most cases sporadic.

199

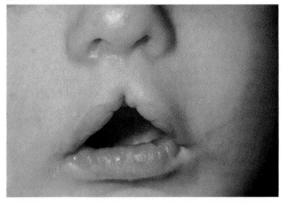

200

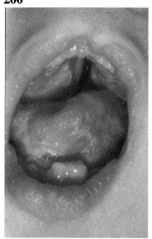

201

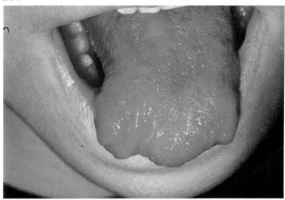

202

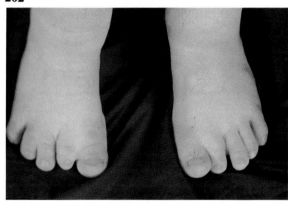

199–202 Oro-facio-digital syndrome type 1.
Note: Mid-line cleft of upper lip, cleft of hard and
soft palate, multiple oral frenulae, shortening and
syndactyly of toes and fingers.

Other features: Mental retardation, seizures,
polydactyly and renal anomalies.

Inheritance: X-linked dominant. Hemizygous
males probably die *in utero*.

203

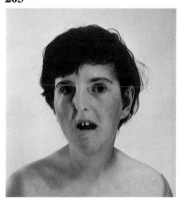

204

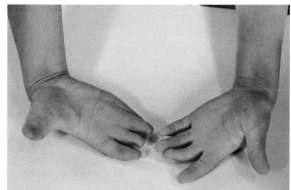

N.B. **203–206.**
Oro-facio-digital syndrome type 2 (Mohr syndrome) is distinguished by autosomal recessive
inheritance, the presence of conductive deafness,
pre-axial polydactyly and fatty tumours of the
tongue.

205

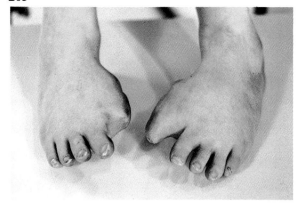

206

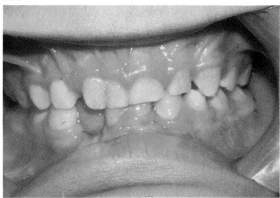

207 HMC syndrome (hypertelorism-microtia-clefting).
Note: Hypertelorism, facial clefting and small dysplastic ears.
   Other features: Congenital heart defects, renal anomalies.
   Inheritance: Possibly autosomal recessive.

207

208

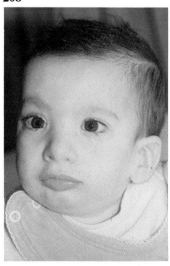

209

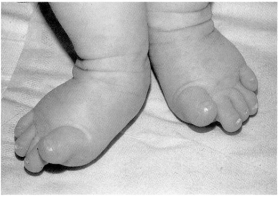

208–211   Coffin–Siris syndrome.
Note: Wide mouth, thickish lips, mild microcephaly, hypoplastic finger and toe nails and hypoplastic terminal phalanges (especially 5th).
   Other features: Mental retardation, sparse scalp hair.
   Inheritance: Autosomal recessive.

210

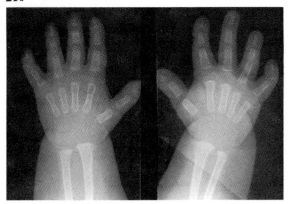

211

212   Opitz C-syndrome.

Note: Prominent metopic suture, upslanting palpebral fissures, small nose and strabismus.

Other features: Contractures, polysyndactyly, dysplastic ears, loose skin, aberrant oral frenulae and mental retardation.

Inheritance: Autosomal recessive.

212

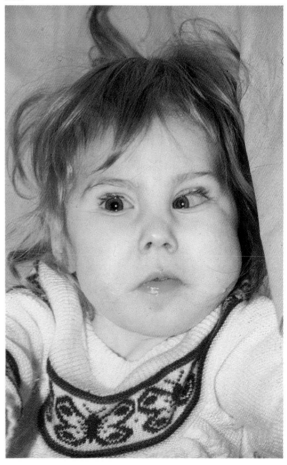

213

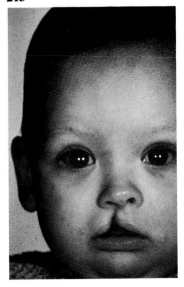

214

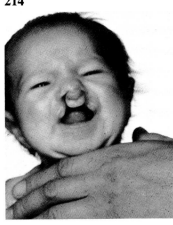

215

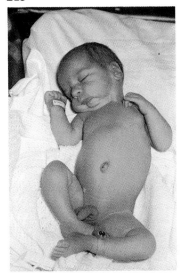

215 Pierre Robin sequence.
Note: Micrognathia.
  Other features: Cleft palate, glossoptosis and respiratory difficulties. Pierre Robin sequence can be part of many syndromes.
  Inheritance: Recurrence risk for an isolated case is in the order of 2%.

213 and 214 Cleft lip and palate.
Note: Unilateral cleft lip, bilateral cleft lip.
  Other features: Isolated cleft palate appears to be a separate entity, breeding true in families. Cleft lip with or without cleft palate can be part of many syndromes, which must be excluded.
  Inheritance: Thought to be multifactorial, the recurrence risks are given in the table.

Isolated cleft palate

Risk to sibs and offspring of a single case	2%

Cleft lip and palate

Risk to sibs or offspring of a unilateral case	4%
Risk to sibs or offspring of a bilateral case	5.5%

216

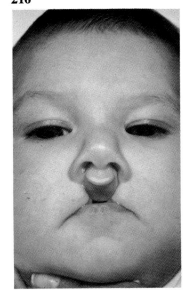

216 Van der Woude syndrome.
Note: Cleft lip and palate, pits on lower lip.
  Inheritance: Autosomal dominant.

217

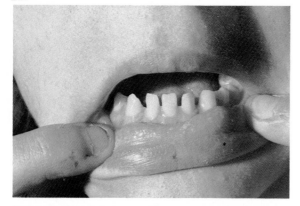

218

217 and 218   Rieger syndrome.
Note: Small, peg-like teeth, irregular placement and oligodontia. Dysplastic iris with irregular outline of the pupil.
   Other features: Glaucoma, corneal opacities, ectopia lentis and occasional mental retardation.
   Inheritance: Autosomal dominant.

219

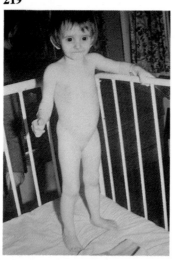

220

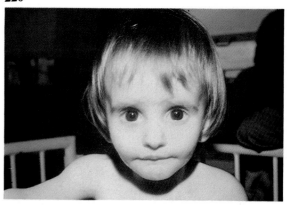

219 and 220   Russell–Silver syndrome.
Note: Small triangular face, prominent forehead giving 'pseudo-hydrocephalus' appearance and thin lips with tendency to downturned corners.
   Asymmetry of limbs with proportionate short stature.
   Other features: Clinodactyly, *café au lait* spots, prenatal growth deficiency and generally normal intelligence.
   Inheritance: Usually sporadic.

221

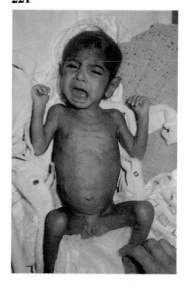

221   Leprechaunism (Donahue syndrome).
Note: Emaciated appearance, hirsutism, thickish lips, wide nostrils and large clitoris.
   Other features: Prominent eyes, mental retardation, low birth-weight, failure to thrive and hyperinsulinism.
   Inheritance: Autosomal recessive.

222

223

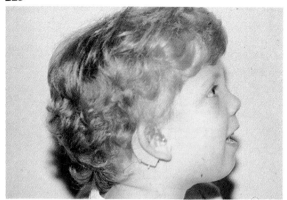

224

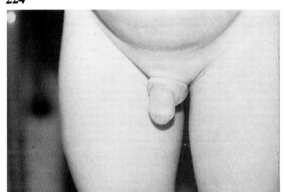

225

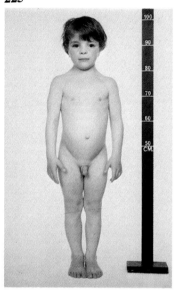

226

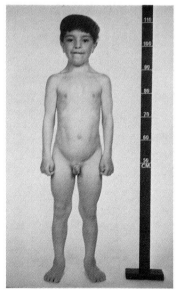

222–226 Aarskog syndrome.

Note: Hypertelorism, anteverted nostrils, small nose, broad philtrum, maxillary hypoplasia, 'shawl' scrotum, lax joints, bulbous tips to digits and clinodactyly.

Other features: Mild ptosis, short stature.

Inheritance: X-linked recessive, carrier females sometimes show mild manifestations and especially short stature.

227 Seckel ('Bird-headed dwarf').

Note: Microcephaly, small face with prominent nose.

Other features: Marked growth deficiency of prenatal origin, mental retardation, clinodactyly of 5th finger and minor hand anomalies, dislocation of hip and dental abnormalities.

Inheritance: Autosomal recessive.

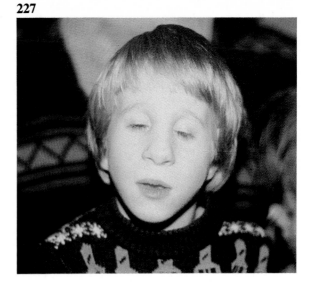

227

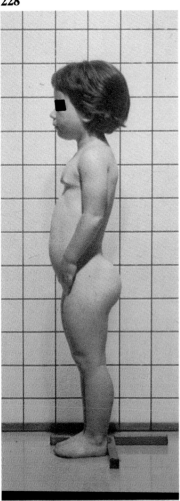

228

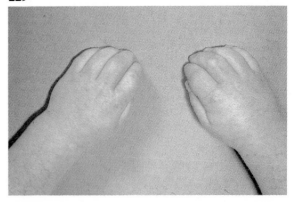

229

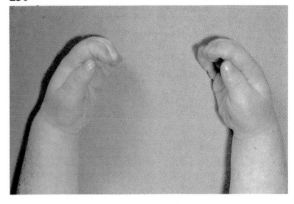

230

228–230 Moore–Federman dwarfism.

Note: Short stature, lumbar lordosis.

Brachydactyly, subcutaneous swelling and inability to form fist.

Other features: Hypermetropia, stiff joints and normal intelligence.

Inheritance: Autosomal dominant.

231

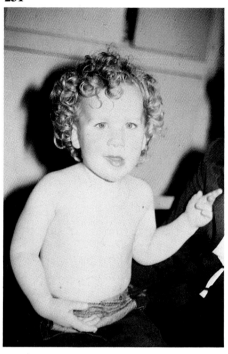

231–233 Sotos syndrome.
Note: Macrocephaly with high, broad forehead, downslanting palpebral fissures, prognathism with narrow chin and high palate.

Other features; Increased growth rate of prenatal onset.

Inheritance: Mild mental retardation, large hands and feet. Mostly sporadic. Occasional dominant pedigrees described.

232

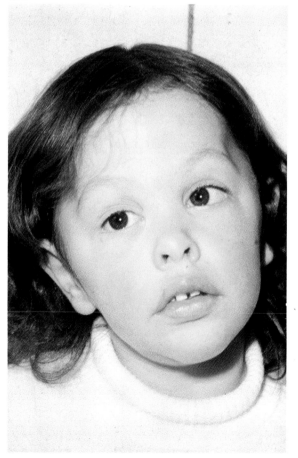

233

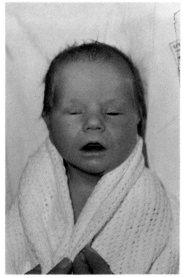

234

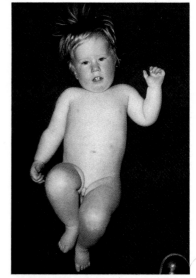

235

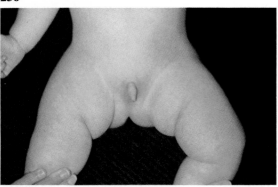

236

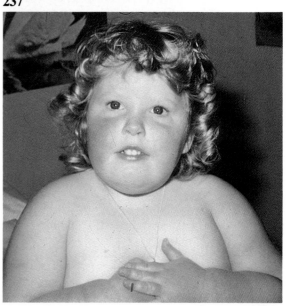

237

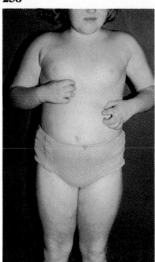

238

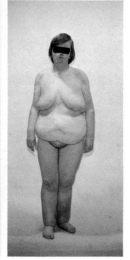

239

234–239 Prader–Willi syndrome.
Note: Neonatal hypotonia with obesity of face, trunk and limbs developing in the first and second year of life. Prominent forehead, almond-shaped eyes, triangular-shaped upper lip.

Other features: Mental retardation and short stature with small hands and feet. Tendency to develop diabetes.

Inheritance: Usually sporadic. There is an association with chromosome re-arrangements and deletions involving the long arm of chromosome 15.

240

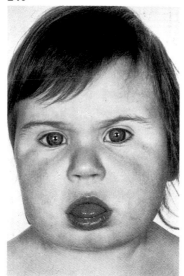

241

242

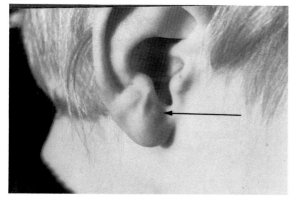

243

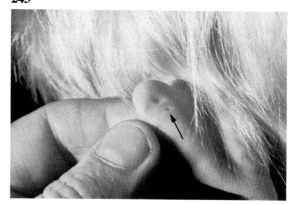

240–243 Beckwith–Wiedemann syndrome.
Note: Large tongue with open mouth. Infra-orbital hypoplasia. Horizontal creases on the lobe (arrow) of the ear together with small, punched-out pits behind the helix (arrow).

Other features: Accelerated growth and osseous maturation, omphalocele, organomegaly, diaphragmatic eventration, pancreatic islet cell hyperplasia leading to hypoglycaemia.

Inheritance: Mostly sporadic, but occasional dominant pedigrees described.

244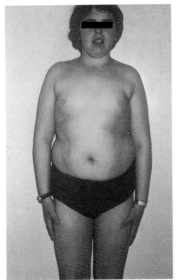

245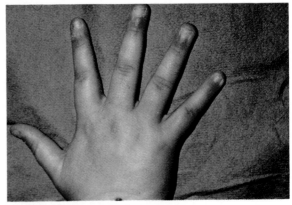

246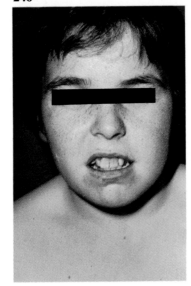

244–246 Cohen syndrome.
Note: Truncal obesity, prominent incisors and open mouth.
Long thin fingers.
Other features: Mental retardation, scoliosis, delayed puberty, ante-mongoloid slant to eyes and eye anomalies.
Inheritance: Probably autosomal recessive.

247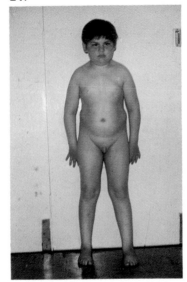

247 Börjeson–Forssman–Lehmann syndrome.
Note: Obesity, gynaecomastia, hypogenitalism, fatty face and cheeks giving a narrow appearance to the palpebral fissures, large ears and genu valgum.
Other features: Mental retardation, seizures.
Inheritance: X-linked recessive.

248

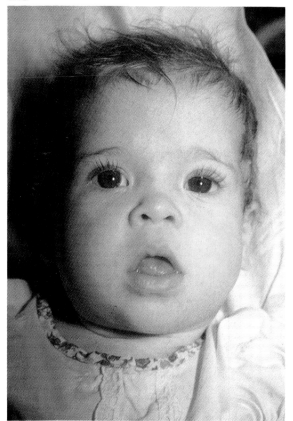

249

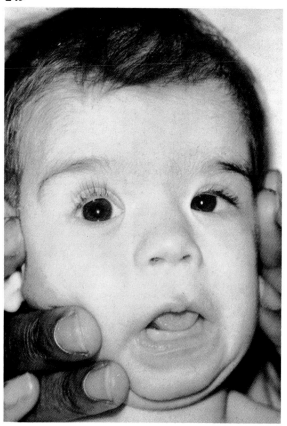

248–250 Fetal hydantoin syndrome.

Note: Mid-face hypoplasia, hirsute forehead, prominent epicanthus, depressed nasal bridge, broad alveolar ridge and hypoplastic nails.

Other features: Mild growth and mental retardation, cleft lip and palate, low-set ears, microcephaly and congenital heart defects.

Aetiology: Maternal epilepsy plus hydantoin therapy. The risk of a child with the full syndrome is 5–10% in mothers taking hydantoins in the early part of pregnancy.

250

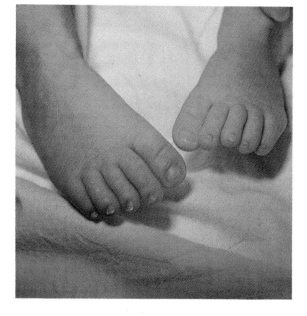

251

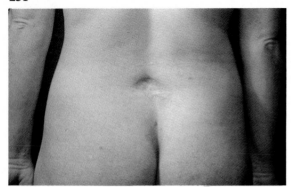

252

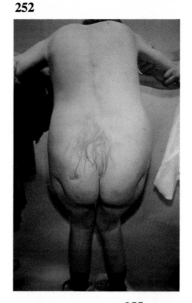

253 **254** **255**

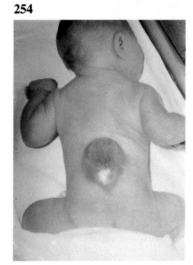

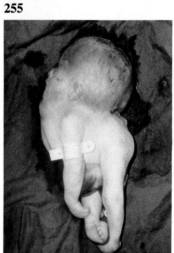

251–257 Neural tube defects.

Note:

(a) Spina bifida occulta with depressed, hairy patch on back.

(b) Meningo-myelocele, large cystic swelling in thoracolumbar region containing nervous tissue.

(c) Iniencephaly, absence of neck with severe retroflexion of head due to gross disruption of vertebral column and meningo-myelocele, exomphthalos.

(d) Encephalocele, microcephaly with occipital encephalocele.

(e) Anencephaly, absence of cranium.

Inheritance: 'Multifactorial'; recurrence after one affected child – 1 in 25; after two 1 in 10 for any neural tube defect.

N.B. Risk applies to sibs and offspring. Prenatal diagnosis is possible by estimating amniotic fluid alpha-feto-protein (AFP) levels and by detailed ultrasound. Maternal serum AFP estimation can be used as a screening test during pregnancy. Ninety per cent of anencephalic and 80% of spina bifida cases will be detected.

256

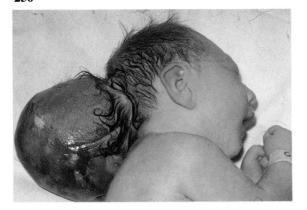

257

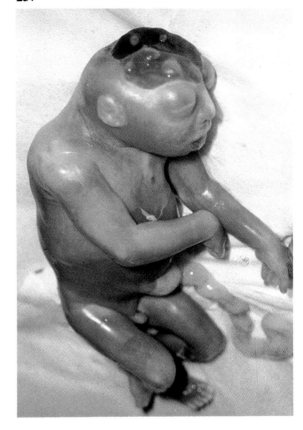

258

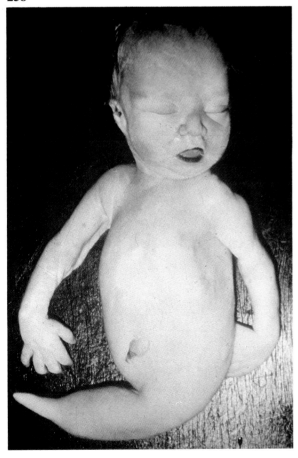

259

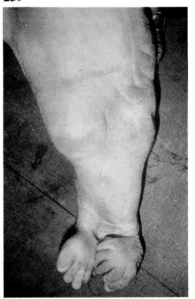

258 and 259   Sirenomelia.
Note: 'Potters' facies, complete fusion of lower limbs with absence of genitalia.

Other features: Vertebral defects, renal agenesis and imperforate anus.

Inheritance: Sporadic, more common in one of identical twins.

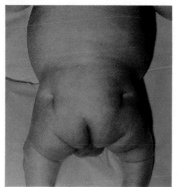

260

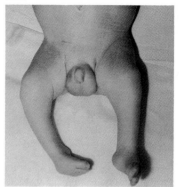

261

260 and 261   Diabetic embryopathy.
Note: Gross muscle wasting of legs, severe talipes equinovarus, short femora and sacral agenesis with prominent dimpling of skin.

   Other features: Heart defect, cleft palate. In general infants of diabetic mothers have 2–3 times the incidence of congenital malformations.

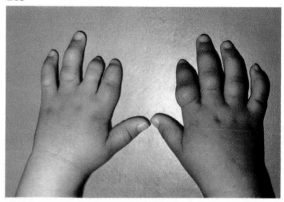

263

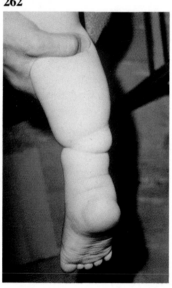

262

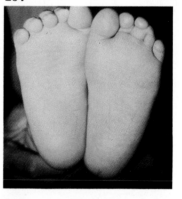

264

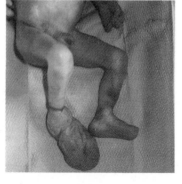

265

262–265   Amniotic bands.
Note:
(a) Constriction of skin and subcutaneous tissue of the calf.
(b) Partial amputation and soft tissue constriction rings affecting fingers bilaterally.
(c) Partial syndactyly of toes and constriction bands.
(d) Amputation of toes on left, constriction band on the right with gross distal swelling.

   Inheritance: Mostly sporadic.

266   Klippel–Feil anomaly.
Note: Short neck with elevated scapulae (more pronounced on the right than left), scoliosis.

Other features: Low hair-line, fusion of cervical vertebrae and other vertebral anomalies, increased incidence of deafness and congenital heart defect.

Inheritance: Mainly sporadic.

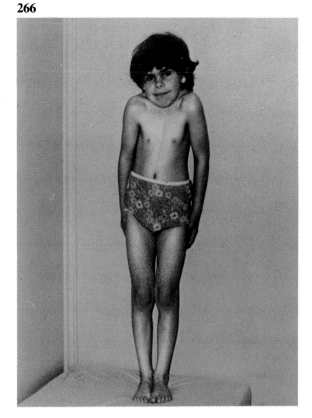

267

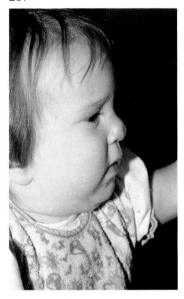

268

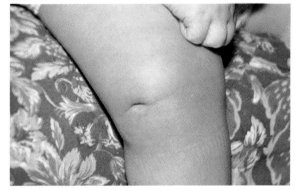

267 and 268   Fetal compression effects.
Note: Small jaw and depressed nasal tip caused by a persistent *in utero* transverse lie. Deep skin depression over knee, indicating fetal constraint (the patient also has widespread joint contractures).

Other features: Fetal compression can cause a wide variety of deformations such as craniostenosis, micrognathia with cleft palate and positional deformities of the limbs. Uterine malformations in the mother are a common association.

8 Bone dysplasias

269–271 Thanatophoric dysplasia.

Note: Body – short bowed limbs, small thorax, large head and mid-face hypoplasia with anteverted nostrils.

Other features: Occasional 'clover-leaf' skull anomaly.

X-rays: Very short tubular bones with metaphyseal flaring and cupping, short, 'telephone receiver', femora with medial metaphyseal spurs, flat vertebral bodies giving inverted – 'U' or 'H'-shaped appearance, in the A-P projection.

Inheritance: All well-documented cases have been sporadic.

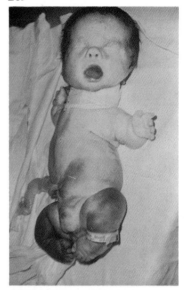

269

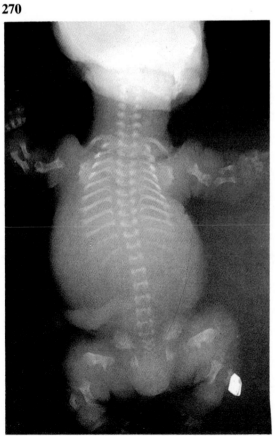

270

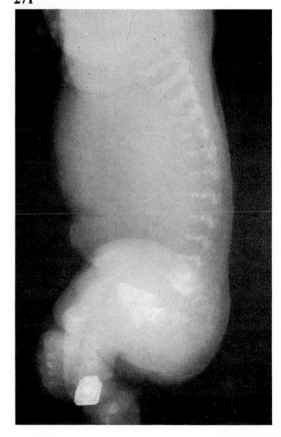

271

272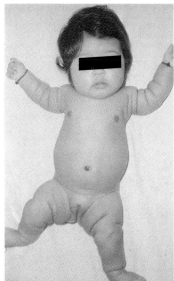

273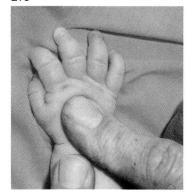

274

275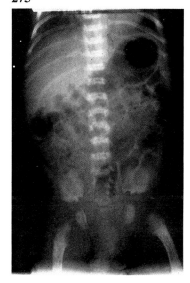

276

272–276 Achondroplasia.

Note: Photo – short stature, large head, prominent forehead, mid-face hypoplasia, genu varum, 'trident' hands and lumbar lordosis.

X-rays: Decreasing interpeduncular distance from thoracic to lumbar spine, short round iliac crests, narrow sacro-sciatic notches, horizontal acetabular roof and oval translucency of proximal femora.

Inheritance: Autosomal dominant, about 70–80% of cases are new mutations.

277

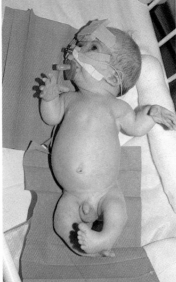

278

279

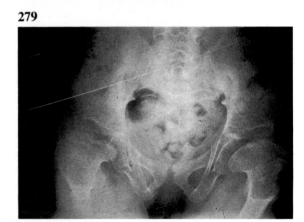

277–279   Diastrophic dysplasia.
Note: Hand – proximally inserted, short and abducted thumb (Hitchhiker's thumb).

Body – short limbs, micrognathia and club feet.

Other features: Cleft palate, joint contractures, kyphoscoliosis and cystic swelling of the ears.

X-ray features: Note – wide metaphyses with epiphyseal flattening, broad femoral necks, narrow interpeduncular distances, in an adult.

Other features: Shortening and metaphyseal widening of tubular bones, short first metacarpal, flattened and irregular vertebral bodies.

Inheritance: Autosomal recessive.

280

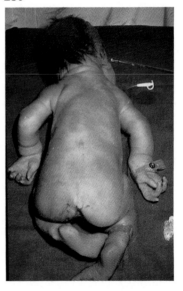

280   Metatropic dysplasia.
Note: Narrow chest, short limbs, prominent joints, long feet and caudal appendage.

Other features: Short stature, severe kyphosis and scoliosis and enlarged joints. X-rays show extreme flaring of the metaphyses in the neonatal period with platyspondyly.

Inheritance: Autosomal recessive.

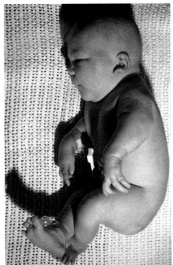

281

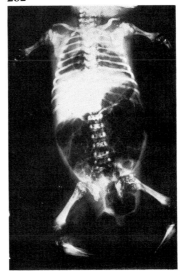

282

281 and 282   Chondrodysplasia punctata, recessive type.

Photo: Note – shortening of limbs, most severe in the upper segments, flat mid-face and dry skin with some dimpling.

Other features: Ichthyosiform skin changes, cataracts. Mental retardation and death in infancy.

X-rays: Note – severe shortening of proximal segments, expanded metaphyses, epiphyseal and extra-cartilagenous stippling.

Other features: Prominent coronal clefts .of lumbar vertebrae.

Inheritance: Autosomal recessive. A milder form with asymmetric shortening of the limbs is autosomal dominant.

283

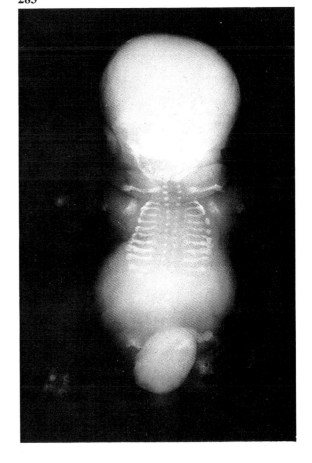

283 and 284   Achondrogenesis type I.

Note: Deficient ossification of vertebral bodies pelvis and sacrum, multiple fractures and under-ossification of the ribs and severely shortened long bones with spur formation.

Inheritance: Autosomal recessive.

284

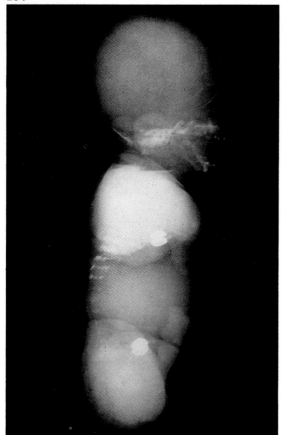

285

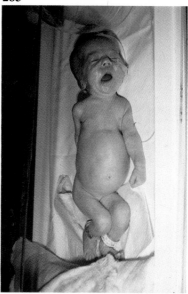

285 and 286 Asphyxiating thoracic dystrophy (Jeune).

Note: Small and narrow thoracic cage. Short ribs, 'trident' configuration to acetabular roof.

Other features: Polydactyly (occasionally), cystic renal dysplasia, hepatic fibrosis and normal intelligence if patient survives.

Inheritance: Autosomal recessive.

286

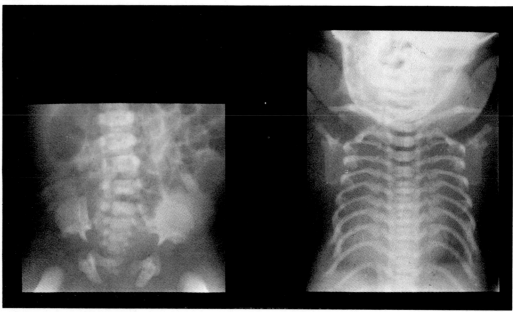

287 Hypophosphatasia – congenital type.

Note: Short limbs, gross undermineralization of all bones, splayed and cupped metaphyses of long bones, short, thin ribs and poor ossification of sternum.

Other features: Death at, or within a few hours of birth, gross undermineralization of the skull, low serum alkaline phosphatase, increased excretion of phospho-ethanolamine and craniostenosis.

Inheritance: Autosomal recessive. Antenatal diagnosis possible.

N.B. There is also a dominant form of the condition with milder features. In addition, some recessive pedigrees show milder features, compatible with a normal life span.

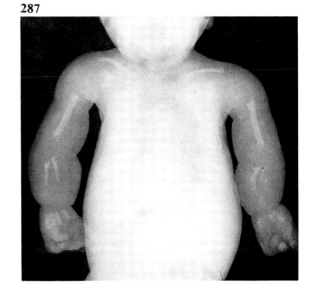

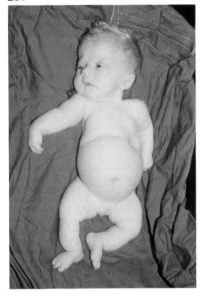

288 and 289 Osteogenesis imperfecta – congenital lethal type.

Note: Bowing of limbs, narrow deformed chest, beaked nose and high forehead.

X-rays: Note – gross shortening of long bones ('crumpled' appearance) with multiple fractures. The ribs are beaded due to multiple fractures *in utero.*

Other features: Wormian bones in the skull.

Inheritance: Some well-documented autosomal recessive sibships have been reported, but many cases are sporadic. A recurrence risk of 1 in 20 after a sporadic case is appropriate.

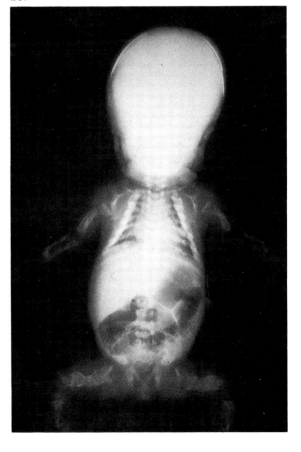

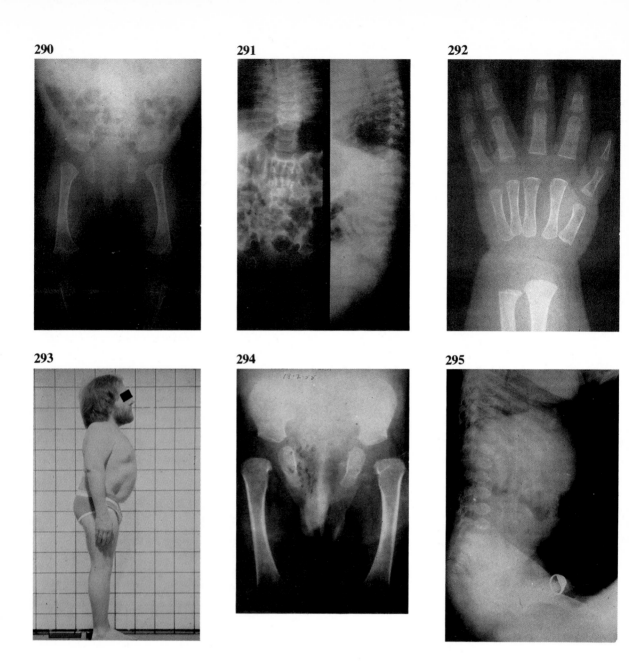

290–295 Spondylo-epiphyseal dysplasia congenita.

Note:

(a) Short spine and lordosis in an adult.

(b) Absence of ossification of pubis and femoral heads, short iliac bones with horizontal acetabular roofs.

(c) Oval or pear-shaped vertebral bodies, anterior pointing at thoracolumbar level (appearance in childhood).

Other features: Cleft palate, myopia, retinal detachment.

Inheritance: Autosomal dominant.

296 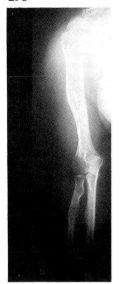 **297** 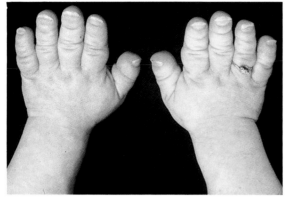 **298**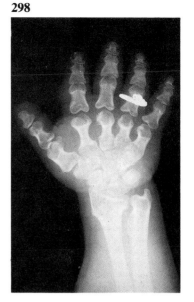

299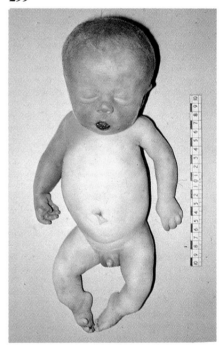

296–298 Acro-mesomelic dwarfism.

Note: Shortening of forearm with markedly short and stubby digits.

X-ray features: Short, broad phalanges with cone-shaped epiphyses, short bowed radius with dorsal subluxation of radial head and oval vertebral bodies with anterior protrusion.

Other features: Short stature, broad, tall forehead with short stubby nose and normal intelligence.

Inheritance: Autosomal recessive.

300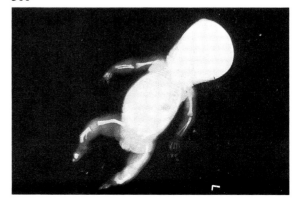

299 and 300 Campomelic dysplasia.

Photo: Note – short, bowed limbs, talipes equinovarus, flat mid-face, low nasal bridge, micrognathia and ambiguous genitalia.

X-rays: Note – bowing of long bones, hypoplastic scapulae.

Other features: 'Clover-leaf' skull anomaly occasionally, sex-reversal.

Inheritance: Evidence of heterogeneity with short and long-bone types. Some cases autosomal recessive.

301

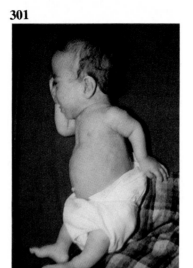

302

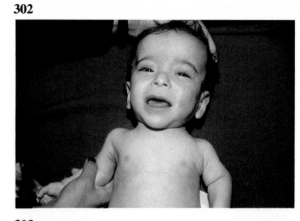

303

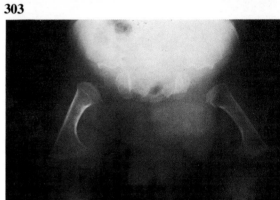

304

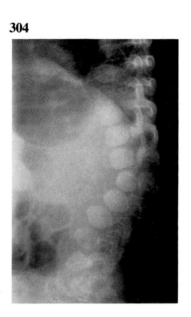

305

301–305 Pseudo-achondroplasia.

Photo: Note – dwarfism with short, bowed limbs and relatively normal cranio-facial appearance.

X-rays: Note – flattening of vertebral bodies with biconvex appearance, horizontal acetabulae with irregularity of articular surface, shortened femora with expanded metaphyses, delayed ossification of femoral and capital epiphyses.

Other features: Small, irregular epiphyses and kyphoscoliosis.

Inheritance: Considerable heterogeneity with autosomal recessive and dominant types.

306 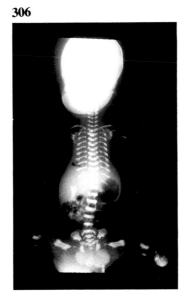 **307**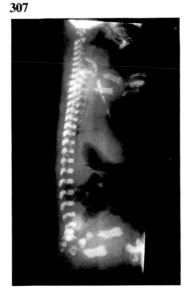

306 and 307 Short-rib polydactyly type I (Saldino–Noonan).

Note: Severely shortened ribs, small pelvis and very short tubular bones with jagged appearance at the metaphyses.

Other features: Polydactyly, hydropic appearance at birth and early death.

Inheritance: Autosomal recessive.

308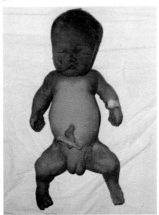

309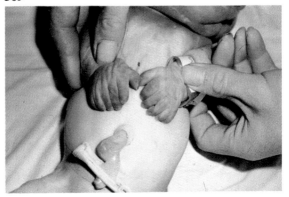

310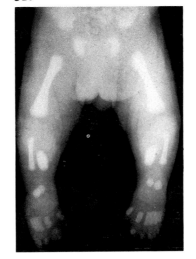

308–310 Short-rib polydactyly type II (Majewski).

Note: Hands – post-axial polydactyly.

Body – narrow thorax, short limbs, short flat nose with median cleft lip, ambiguous genitalia and slight hydropic appearance.

X-rays: Relatively normal pelvis, short oval tibiae and polydactyly of feet.

Other features: Short, horizontal ribs and poly-syndactyly of hands.

Inheritance: Probably autosomal recessive.

311 Hypochondroplasia.

Note: Short stature, increased lumbar lordosis, rhizomelic shortening of limbs and mild frontal bossing.

Other features: Mental retardation sometimes a feature.

X-ray features: Some X-ray features of achondroplasia, such as short broad femoral neck, moderate narrowing of interpeduncular distance in lumbar region and small greater sciatic notches.

Inheritance: Autosomal dominant.

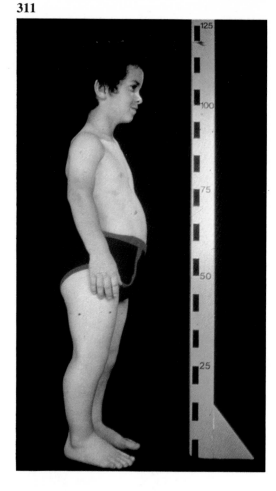

311

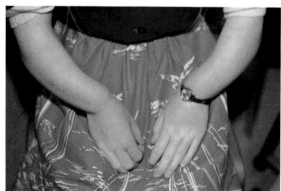

312

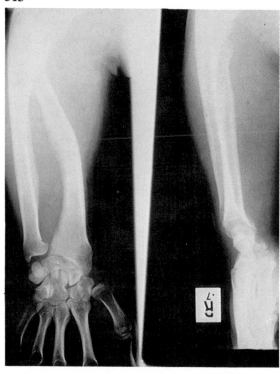

313

312 and 313 Dyschondrosteosis.

X-rays: Note – dorsal subluxation of radius and ulna, bowing of radius with 'V'-shape configuration to distal end of radius and ulna.

Photo: Note – bilateral Madelung's deformity at the wrist, and short forearms.

Inheritance: Autosomal dominant. Homozygotes present with severe mesomelic dysplasia (Langer type).

314   Spondylo-epiphyseal dysplasia tarda.
Note: Flattening of vertebrae with 'hump-shaped' build-up of bone on superior and inferior surfaces.

Other features: Short stature, irregular epiphyses.

Inheritance: X-linked recessive, occasional dominant families.

314

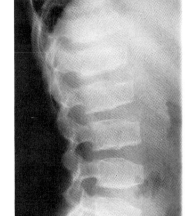

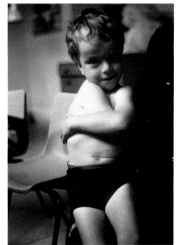

315–317   Cleido-cranial dysplasia.
Note: Absence of clavicle allowing approximation of shoulders, frontal bossing.

Other features: Brachycephaly, mid-face hypoplasia, large anterior fontanelle and late eruption of teeth.

X-ray skull: Note – persistent anterior fontanelle and multiple Wormian bones.

Chest: Note – very small, hypoplastic clavicles.

Inheritance: Autosomal dominant.

316

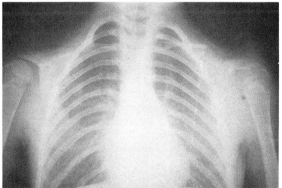

317

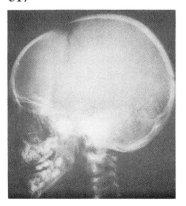

318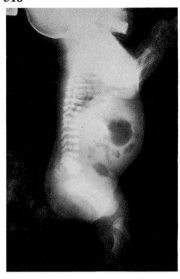
319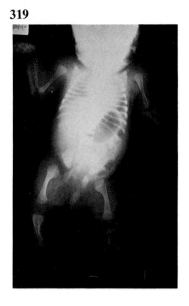
320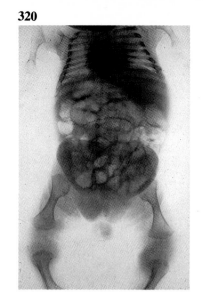

321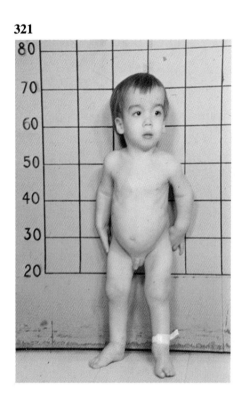

322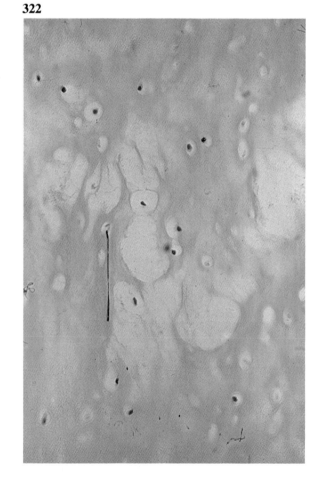

318–322   Kniest dysplasia.

Note: Short 'barrel-shaped' chest, expanded metaphyses and poor ossification of epiphyses.

Coronal clefts of lumbar vertebrae and platyspondyly.

Flat mid-face, short trunk and prominent joints.

Other features: Cleft palate, deafness and retinal detachment. Histopathology of cartilage shows 'Swiss-cheese' pattern.

Inheritance: Autosomal dominant.

323

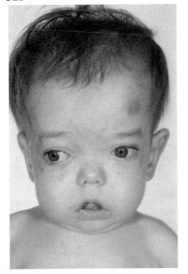

324

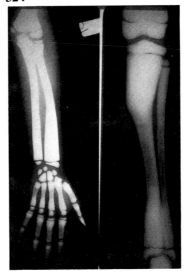

323 and 324 Osteopetrosis.
X-rays: Note – increased density of long bones with defective modelling.
  Other features: 'Bone within bone' appearance.
  Face: Note – facial weakness, broad forehead and prominent jaw.
  Other features: Bone fragility, optic atrophy, deafness and leuco-erythroblastic anaemia.

Inheritance: The severe, precocious form is autosomal recessive. The more benign form with late manifestations is usually autosomal dominant, although autosomal recessive families have been reported.

325

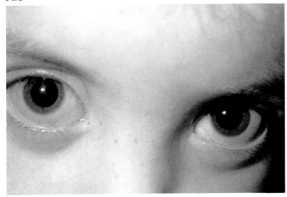

326

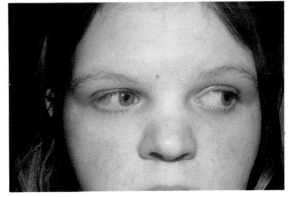

327

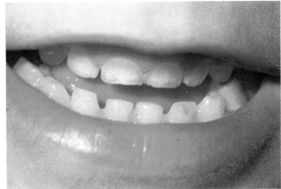

325–327 Osteogenesis imperfecta-tarda.
Face: Blue sclerae.
  Yellow, opalescent teeth – dentinogenesis imperfecta.
  Other features: Multiple fractures, deafness, loose joints and thin skin.
  X-ray features: Demineralized bones with thin cortex, evidence of old and new fractures and Wormian bones in the skull.
  Inheritance: Mostly autosomal dominant.

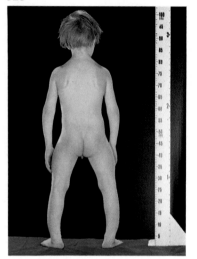

328

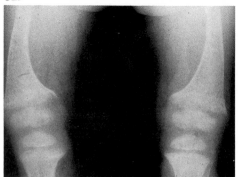

329

328 and 329 Hypophosphataemic rickets.
X-ray: Note – metaphyseal widening with splayed, irregular and cupped appearance, rarefaction, coarse trabeculation and curvature of femora.

Other features: Curvature of long bones, 'pseudo-fractures', scoliosis, low serum phosphate, moderate elevation of alkaline phosphatase and hyperphosphaturia.

Inheritance: X-linked dominant.

330

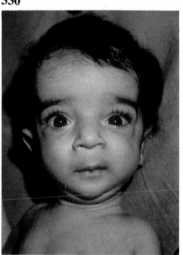

331

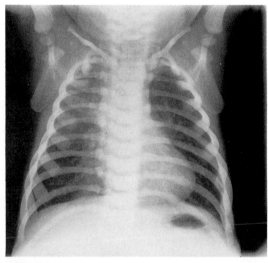

332

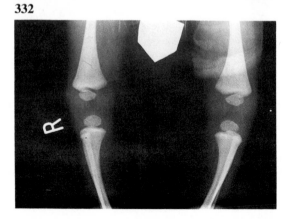

330–332 Pyknodysostosis.
Note: Prominent forehead, small chin with triangular face, prominent eyes and bluish sclerae. Dense bones on X-ray, slightly hypoplastic clavicles.

Other features: Large anterior fontanelle, tendency to fractures, generalized osteopetrosis, Wormian bones, increased angle of mandible, acro-osteolysis and short stature.

Inheritance: Autosomal recessive.

333

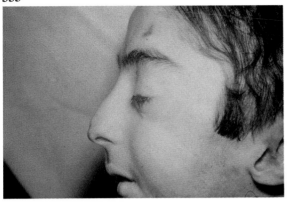

334

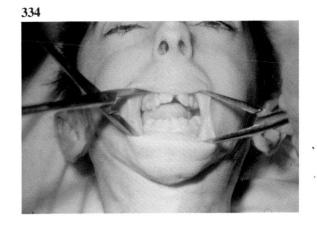

335

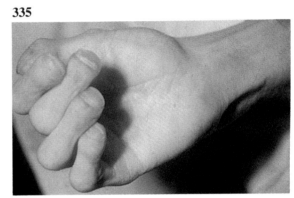

333–335 Mandibulo-acral dysplasia.
Note:
(a) Hooked thin nose, senile appearance.
(b) Small mandible, crowded teeth (post-mortem photograph).
(c) Shortened and expanded tips to fingers, atrophic skin.
  Other features: Kyphoscoliosis, clavicular hypoplasia and osteolysis of terminal phalanges.
  Inheritance: Autosomal recessive.

336

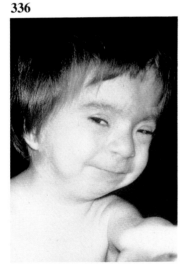

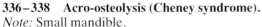

337

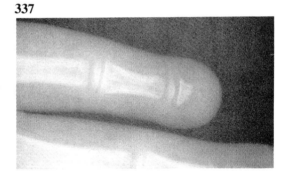

338

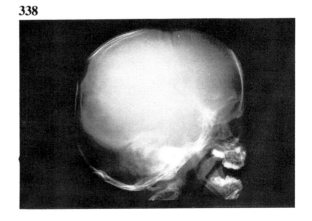

336–338 Acro-osteolysis (Cheney syndrome).
Note: Small mandible.
  X-rays: Osteolysis of terminal phalanges, Wormian bones, wide sutures, absence of frontal sinuses, elongated sella turcica and relatively normal mandibular angle.
  Other features: Short stature, early loss of teeth and kyphoscoliosis.
  Inheritance: Autosomal dominant.

339

340

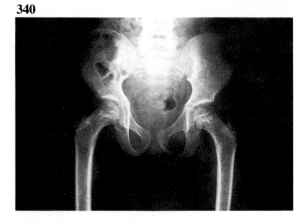

341

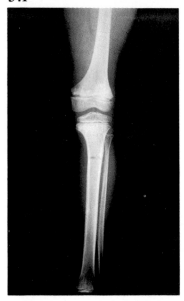

339–341   Cartilage-hair hypoplasia (McKusick).

Note: Fine, sparse hair, short fingers, metaphyseal irregularities and some flattening of the epiphyses. The changes are more severe at the knee than in the proximal femur. The distal fibula is relatively long.

Other features: Short stature, small feet and impaired cellular immunity.

Inheritance: Autosomal recessive.

342

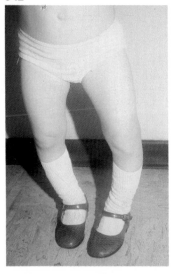

343

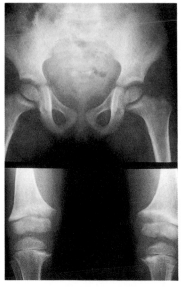

342 and 343   Metaphyseal dysplasia (Schmid).

Note: Early metaphyseal irregularities with cupping of the growth plate.

   Bowed legs, short stature.

Inheritance: Autosomal dominant.

344

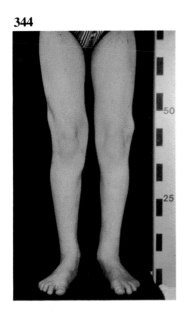

345

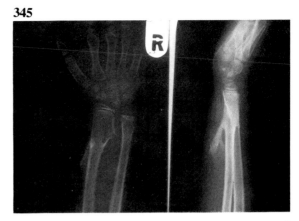

344 and 345 Multiple exostoses (diaphyseal aclasis).
Note: Exostoses of distal end of radius, on X-ray. Bony protrusions around the knee.
Other features: Sarcomatous degeneration in 5–10% of patients.
Inheritance: Autosomal dominant.

346

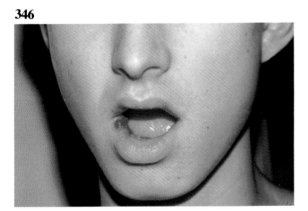

347

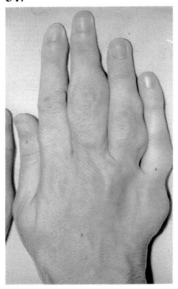

346 and 347 Maffucci syndrome.
Note: Haemangiomatous lesions of buccal mucosa, enchondromata of bones of the hand.
Inheritance: Sporadic.

348   **349**

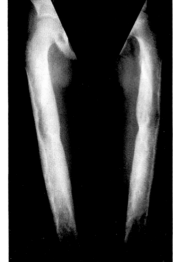

348 and 349 Camurati–Engelmann syndrome (progressive diaphyseal dysplasia).
Note: Thickening of cortices of the long bones with narrowed medullary canals.
Other features: Sclerosis of skull, muscular pain and weakness.
Inheritance: Autosomal dominant with variable penetrance.

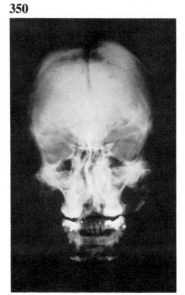

350

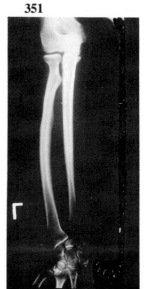

351

350–352 Kenny–Caffey syndrome.
Note: Sclerosis of skull with open anterior fontanelle and metopic suture, slim medullary cavities of long bones.

Other features: Short stature, myopia and hypocalcaemia in infancy.

352

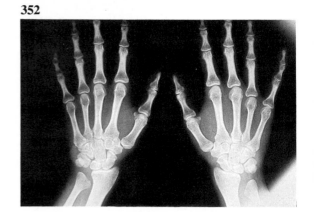

353

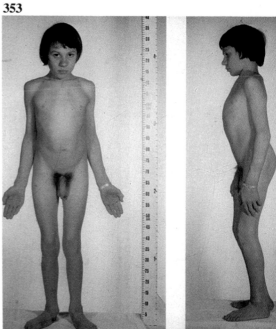

353–355 Dyggve–Melchior–Clausen syndrome.
Note: Short trunk, protruding sternum and lumbar lordosis.

X-rays: Lace-like appearance of iliac crests, hypoplastic basal portion of ilia and epiphyseal irregular end-plates.

Inheritance: Autosomal recessive.

354

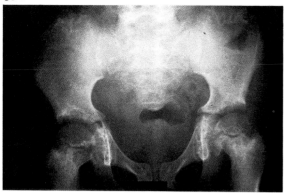

355

356   Jarcho–Levin syndrome.
Note: Multiple vertebral defects consisting of hemivertebrae, abnormal and crowded ribs.

Other features: Unusual facies consisting of anteverted nostrils, broad forehead and upslanting eyes.

Inheritance: Autosomal recessive.

356

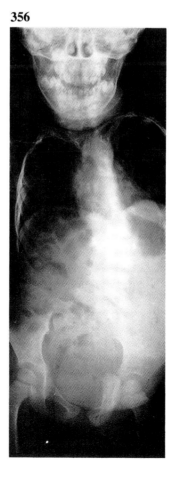

9 Deafness and ear malformations

Deafness may be divided into congenital types, and those developing later in life.

Congenital deafness (prevalence 45 per 100,000)

There are many different causes of congenital deafness. A rough breakdown with percentages is given below:

Causes of congenital deafness	%
Environmental	35
Unknown	20
Genetic	45

Genetic forms can be broken down into roughly the following percentages:

Autosomal recessive	66
Autosomal dominant	31
X-linked recessive	3

Approach to counselling in congenital deafness

(a) Exclude environmental causes in the proband. Possible perinatal causes include:
   (i)   rubella,
   (ii) toxoplasmosis,
   (iii) rhesus incompatibility,
   (iv) severe oxygen deprivation,
   (v) drugs (e.g. gentamicin).
(b) Are there additional physical or clinical features in the proband?
(c) Are there other affected family members?

Where there are no additional clinical features counselling can be demonstrated by a number of hypothetical pedigrees.

357

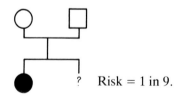

? Risk = 1 in 9.

Reasoning
After exclusion of environmental causes, 45 out of 65 (0.7) cases are genetic. Of genetic causes about 2 out of 3 are autosomal recessive. Therefore the recurrence risks are:

$$\frac{7}{10} \times \frac{2}{3} \times \frac{1}{4} \approx \frac{1}{9}$$

358

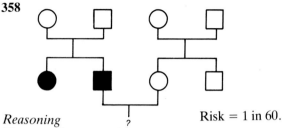

Reasoning ? Risk = 1 in 60.
The pedigree evidence suggests recessive deafness. A carrier frequency in the general population of 1 in 30 is assumed. The recurrence risk is:

$$\frac{1}{30} \times \frac{1}{2} = \frac{1}{60}$$

359

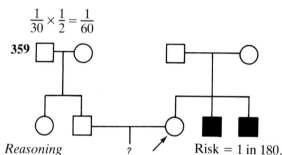

Reasoning ? Risk = 1 in 180.
The pedigree evidence suggests autosomal recessive inheritance. The normal female has a 2 in 3 chance of carrying the gene. Carrier frequency in the population is assumed to be 1 in 30. The recurrence risk is:

$$\frac{2}{3} \times \frac{1}{30} \times \frac{1}{4} = \frac{1}{180}$$

360

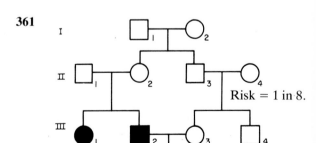

Reasoning ? Risk = 1 in 385.
See above.

$$\frac{7}{10} \times \frac{2}{3} \times \frac{2}{3} \times \frac{1}{30} \times \frac{1}{4} = \frac{1}{385}$$

361

I

II

III

Risk = 1 in 8.

Reasoning
Either I_1 or I_2 could carry the gene. II_3 has a 1 in 2 chance and III_3 a 1 in 4 chance. Therefore risk to offspring of III_2 and III_3 is 1 in 8.

362

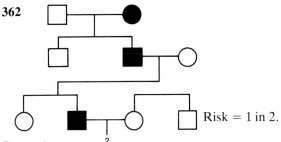

Risk = 1 in 2.

Reasoning
Pedigree suggests autosomal dominant inheritance.

363

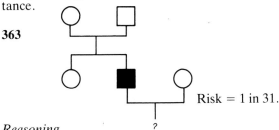

Risk = 1 in 31.

Reasoning
About 5% of *sporadic* cases are dominant. About 45% of *sporadic* cases are recessive. Therefore risk is (approximately):

$$\left[\frac{1}{20} \times \frac{1}{2}\right] + \left[\frac{45}{100} \times \frac{1}{30} \times \frac{1}{2}\right] = \frac{1}{31}$$

364

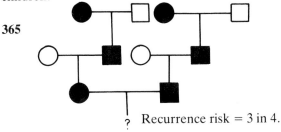

Risk = 1 in 10.

Reasoning
The pedigree suggests autosomal recessive inheritance in both parents. As up to 10 different abnormal recessive genes may be implicated it is therefore possible for parents with different types of autosomal recessive deafness to have normal children.

365

Recurrence risk = 3 in 4.

Reasoning
Dominant × dominant inheritance: if these two deaf people, both of whom have family histories that suggest dominant inheritance, were to reproduce, four genetic combinations would be possible. Three would result in deaf offspring and the fourth in a hearing child.

366

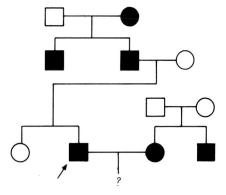

Recurrence risk = 1 in 2.

Reasoning
Dominant × recessive inheritance: if the deaf proband, indicated with an arrow, marries a deaf partner with a recessive type of deafness, the risk of having a deaf child is 1 in 2.

367

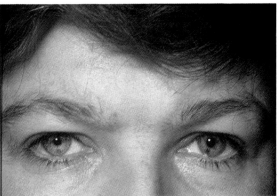

367 Waardenburg syndrome.
Note: Heterochromia of iris, minimal synophrys, dystopia canthorum and white hair in forelock.
   Other features: Deafness in 20–30%. These are possibly two distinct forms with or without dystopia canthorum.
   Inheritance: Autosomal dominant with variable expression.

368

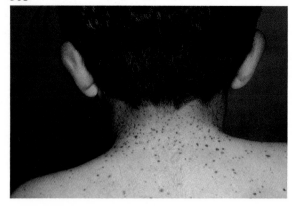

369

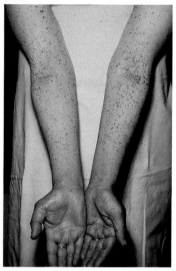

368 and 369   Leopard syndrome.
Note: Lentigenes, more marked on face and upper thorax.

Other features: Lentigenes, electro-cardiographic abnormalities, ocular hypertelorism, pulmonary stenosis, abnormalities of genitalia, retardation of growth and deafness (sensorineural).

Inheritance: Autosomal dominant.

370

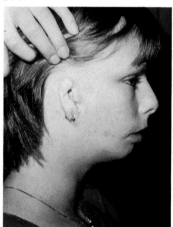

371

370 and 371   Goldenhar syndrome.
Note: Facial asymmetry, hypoplastic zygomatic arch and epibulbar dermoid on left.

Malformed pinna, skin tag removed from right cheek and mandibular hypoplasia.

Other features: Deafness, malformations of heart, kidneys and intestines, vertebral anomalies, coloboma of upper lid and macrostomia. The presence of ocular and other features distinguishes this from the first branchial arch syndrome.

Inheritance: Mostly sporadic but occasional dominant pedigrees described. Appropriate recurrence risk 2%.

372   First branchial arch syndrome.
Note: Severely dysplastic pinna, macrostomia and mandibular hypoplasia.

Other features: Deafness, pre-auricular pits and skin tags.

Inheritance: Usually sporadic.

372

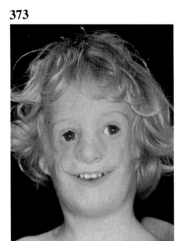
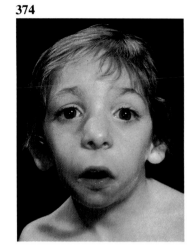

373 and 374 Treacher – Collins syndrome.

Note: Anti-mongoloid slant to eyes, lower lid coloboma, malar hypoplasia, mandibular hypoplasia, partial absence of eyelashes on lower lid and dysplastic ears.

Other features: Deafness, cleft palate.

Inheritance: Autosomal dominant with variable expression.

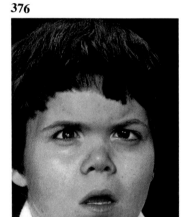

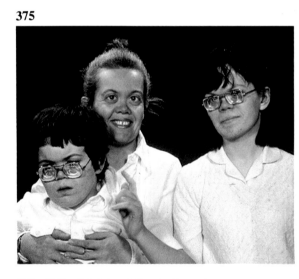

375–377 Stickler–Marshall syndrome.

Note: Mid-face hypoplasia, epicanthic folds and stiffness of joints.

Other features: Deafness, cleft palate, micrognathia, myopia, mild sponylo-epiphyseal dysplasia and normal intelligence. The designation Marshall syndrome is sometimes given when there is unusually prominent mid-face hypoplasia, although the two disorders are probably the same.

Inheritance: Autosomal dominant with variable expression.

378 Branchio-oto dysplasia.

Note: Branchial fistula (arrow).

Other features: Pre-auricular pits, sensorineural deafness. Similar features, together with renal anomalies, is called the branchio-oto-renal syndrome, which is probably a separate entity.

Inheritance: Autosomal dominant with variable expression.

378

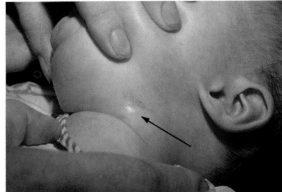

379

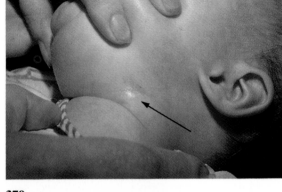

379 Cranio metaphyseal dysplasia.

Note: Prominent frontal bossing, ocular hypertelorism, broad flat nasal bridge and large mandible.

Other features: Mixed hearing loss, facial paralysis, dental malocclusion and widening of metaphyseal areas of long bones.

Inheritance: Dominant and recessive forms noted.

380

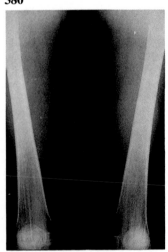

381

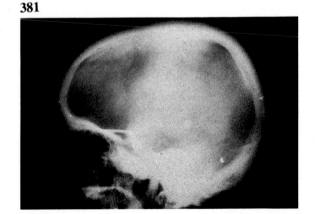

380 and 381 Osteopathia striata with cranial sclerosis.

Note: Linear striation of long bones.

Sclerotic base and vault of the skull.

Other features: Deafness, cleft palate.

Osteopathia striata of the long bones can be a benign, isolated finding.

Inheritance: Autosomal dominant.

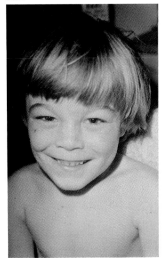

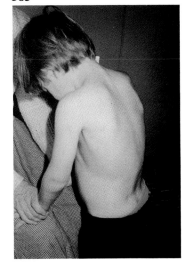

382 and 383   Mannosidosis.

Note: Coarse facial features (prominent supra-orbital ridges, full eyebrows and broad bridge to nose) wide spaced teeth.

Thoracolumbar kyphosis (gibbus).

Other features: Sensorineural deafness, mild mental retardation, hepatosplenomegaly and deficiency of $\propto$-mannosidase.

Inheritance: Autosomal recessive.

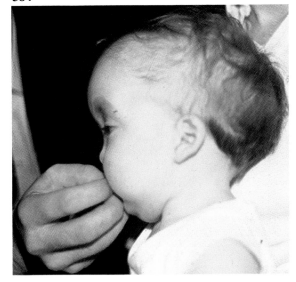

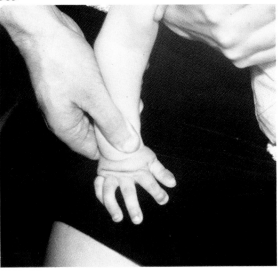

384 and 385   Otopalatodigital syndrome.

Note: Frontal prominence, small nose and mouth and overall 'pugilistic' appearance.

Shortened distal phalanges, abnormal angulation at the interphalangeal joints.

Other features: Similar appearance in the feet giving a 'tree-frog' appearance.

Deafness, hypertelorism, cleft palate and mild mental retardation.

Inheritance: X-linked with mild manifestations in females.

10 Eye disorders

Genetic counselling in blindness and ocular disorders

Prevalence of blindness in childhood is 0.7 in 1000.
The genetic contribution to blindness is:
44% in Europe
8% in developing countries.
The genetic contributions to blindness consists of a large number of syndromes.

A breakdown of the types of inheritance is as follows:

Autosomal dominant	50%
Autosomal recessive	39%
X-linked recessive	11%

There are many different causes of blindness; an approximate distribution is as follows:

Cataract	15%
Glaucoma	7%
Macular degeneration	1%
Retinoblastoma	8%
Optic atrophy	8%
Infections	32%
Other	29%

other instances the type of cataract may give a clue to the inheritance.

Table 1

Type of cataract	Inheritance
Crystalline	A.D. occ. A.R.
Membranous	A.D.
Nuclear	A.D. occ. A.R.
Anterior polar	A.D. or A.R.
Posterior polar	A.D.
Total congenital	A.D. occ. A.R. or X.L.R.
Lamellar (Zonular) with microphthalmos	A.D. A.D., A.R. or X.L.R.

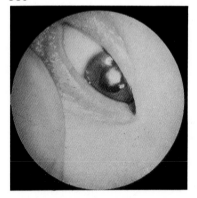

386

386 Congenital cataract.
Note: Large central cataract.
Other features: Cataracts can be part of distinct syndromes (see **Table 1**).
Inheritance: See **Table 1**.

Inheritance of cataract
About 10% of cataracts are thought to be hereditary. Environmental causes and known syndromes must be excluded. Sometimes the type of inheritance can only be inferred from the pedigree. In

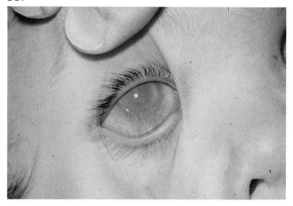

387

387 Corneal dystrophy.
Note: Uniformly opaque cornea.
Other features: Can be part of syndromes (see **Table 2**).
Inheritance: There are five layers to the cornea (epithelium, Bowman's membrane, stroma, Descemet's membrane and endothelium) genetic conditions can affect any of these layers. Autosomal dominant and recessive families have been described. A specialist ophthalmological opinion is needed to differentiate between the types.

Table 2 Genetic syndromes featuring cataracts

Cataracts common (>50%)

Syndrome	Type of cataract	Onset	Inheritance
Galactosaemia	Lamellar	Birth	A.R.
Galactokinase deficiency	Lamellar	Birth	A.R.
Cerebrotendinous xanthomatosis	Lamellar	Childhood	A.R.
Hallerman–Streiff syndrome	Total	Birth	? (Sporadic)
Dystrophia myotonica	Total or stellate	Variable, up to adulthood	A.D.
Lowe syndrome	Variable	Infancy	X.L.R.
Rothmund–Thomson syndrome	Lamellar or total	Infancy	A.R.
Marshall–Stickler syndrome	Lamellar or total	Variable	A.D.
Werner syndrome	Total	Adulthood	A.R.

Cataracts frequent (5–50%)

Syndrome	Type of cataract	Onset	Inheritance
Hypoparathyroidism (and pseudohypoparathyroidism)	Lamellar	Childhood	X-linked or sporadic
Alport syndrome	Lamellar	Variable	? A.D.
Chondrodysplasia punctata (Rhizomelic)	Total	Infancy	A.R.
Cockayne	Variable	Childhood	A.R.
Cerebro-oculo-facio-cerebral syndrome (COFS)	Total	Childhood	A.R.
Incontinentia pigmenti	Variable	Infancy	X.L.D. (?lethal in males)

Table 3 Syndromes featuring corneal opacities

Mucopolysaccharidoses (MPS)	Other
MPS I–H (Hurler)	Fabry disease
MPS I–S (Scheie)	Lipodystrophy (Berardinelli)
MPS IV (Morquio)	Rieger syndrome
MPS VI (Maroteaux–Lamy)	Roberts syndrome

Mucolipidoses (MLS)
MLS III   (PseudoHurler)
MLS IV

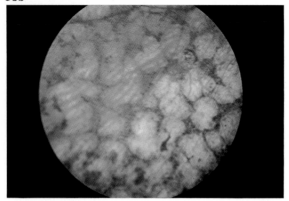

388 Retinitis pigmentosa.
Note: 'Bone corpuscle' pigmentation at the peripheral part of the retina.

Other features: Night blindness, tunnel vision. Atypical retinal pigmentation can be part of syndromes (see **Table 4**).

Genetics of retinitis pigmentosa
Retinitis pigmentosa can be inherited in either an autosomal dominant, recessive or X-linked recessive manner. It is sometimes difficult to differentiate between the genetic types clinically.
(a) *Autosomal recessive.* This is apparently the most common form (>50%). Onset is in the first two decades. Severe visual loss is apparent by the fifth decade.
(b) *Autosomal dominant.* Onset may be similar to the autosomal recessive form, but progression is slower.
(c) *X-linked recessive.* This is probably the least common form. Carrier females can sometimes be diagnosed by fundoscopy and ERG studies.

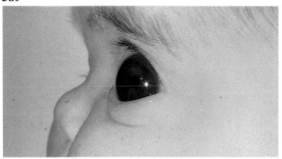

389 Congenital glaucoma (Buphthalmos).
Note: Macrocornea and enlarged globe.

Other features: Cupping of optic discs, blindness if untreated. Can be part of syndromes, Stickler/Marshall, Weill–Marchesani, Lowe and Sturge–Weber.

Inheritance: Autosomal dominant and recessive forms. Recurrence risk for sibs and offspring of an isolated case = 4–5%.

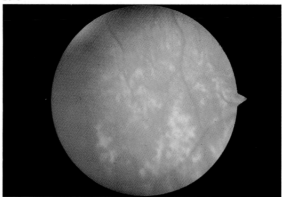

390 Leber amaurosis (congenital retinal blindness).
Note: Attenuation of retinal vessels, granular pigmentation.

Other features: Blindness, absent or marked reduction of ERG and mental retardation in about 50% of cases.

Inheritance: Autosomal recessive; *N.B.* this condition is not the same as Leber optic atrophy (q.v.).

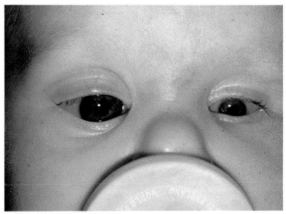

391 Microphthalmia.
Note: Reduction in size of all ocular structures on the left.

Other features: Colobomatous malformations.

Inheritance: Environmental causes (e.g. CMV, rubella and toxoplasmosis) should be excluded. There are also autosomal dominant and recessive forms. The recurrence risk for sibs of an isolated case may be up to 10%. Microphthalmia associated with colobomatous malformations is usually inherited as an autosomal dominant condition.

Table 4 Genetic syndromes featuring abnormal retinal pigmentation

Syndrome	Other features	Inheritance
Usher syndrome	Deafness	A.R.
Alström syndrome	Deafness, diabetes and obesity	A.R.
Refsum syndrome	Neuropathy, ataxia, deafness, ichthyosis, phytanic acid in urine	A.R.
Kearns–Sayre syndrome	External ophthalmoplegia, cardiac conduction defects, deafness and short stature	Mostly sporadic
Laurence–Moon–Biedl syndrome	Obesity, polydactyly and mental retardation	A.R.
Cockayne syndrome	Cachexia, early ageing, cataracts, retardation and deafness	A.R.
RP with nephronophthisis	Nephronophthisis	A.R.
Abetalipoproteinaemia	Acanthocytes, ataxia and lipoprotein abnormalities	A.R.

392

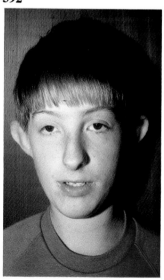

393

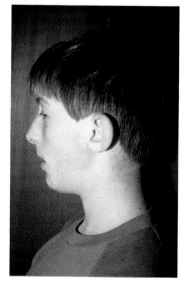

392 and 393   Lenz microphthalmia.
Note: Microphthalmia, simple, protruding ears, thin nose and sloping shoulders.
Other features: Mental retardation, mild short stature, crowded teeth, strabismus and nystagmus, genital anomalies (e.g. hypospadias), camptodactyly and clinodactyly.
Inheritance: X-linked recessive.

394 Leber optic atrophy.

Note: Swelling and pallor of the optic disc with full vessels and retinal oedema.

Other features: Bilateral involvement with acute onset in adolescence. Central scotomata and visual loss.

Inheritance: Male to female ratio 6:1. The mode of inheritance is unlike classic X-linked Mendelian transmission. No descendants of affected males are affected. Risk to offspring of carrier females is more than 50% for affected males or carrier females.

394

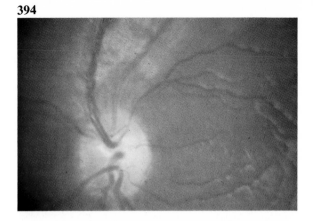

395 Optic atrophy.

Note: Pale optic disc, paucity and attenuation of vessels traversing the disc.

Other features: If dominant, visual loss gradual with variation of severity. If recessive, early onset of severe visual impairment.

Inheritance: Autosomal dominant cases more common than recessive. X-linked pedigrees have very rarely been described.

395

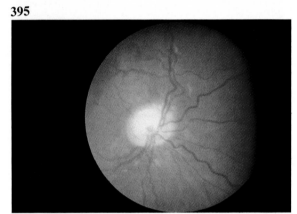

396

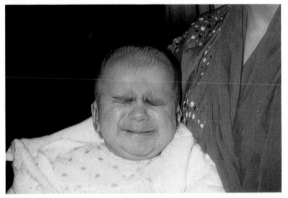

397

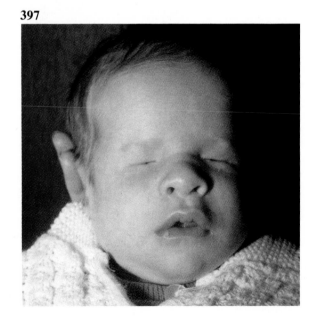

396 and 397   Anophthalmia.

Note: Absent eyes and small orbits.

Other features: Can be part of chromosomal syndromes, e.g. trisomy 13.

Inheritance: Autosomal recessive inheritance common.

398 Aniridia.
Note: Absent irises. Ptosis and photophobia.

Other features: Lens opacities, macular abnormalities and glaucoma. Can be associated with ataxia (Gillespie's syndrome – autosomal recessive) or with Wilm's tumour, genital anomalies and mental retardation (due to a partial deletion of the short arm of chromosome 11).

Inheritance: When isolated, usually autosomal dominant.

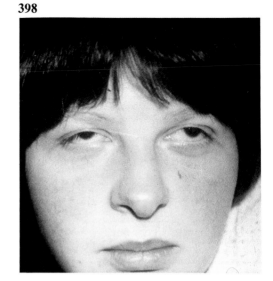

398

399 Retinoblastoma.
Note: Glass eye.

Other features: Ophthalmoscopy shows a white tumour mass with retinal vessels on the surface.

Inheritance: Autosomal dominant pedigrees have been observed (see **Table 5** for risk figures).

N.B. Interstitial deletion of part of the long arm of chromosome 13 predisposes to retinoblastoma.

399

Table 5   Risks in retinoblastoma families

Proband	Risk to offspring	Risk to sibs
Bilaterally affected No family history	50%	2–3%
Unilaterally affected No family history	5–10%	1%

N.B. Parents of isolated cases should be carefully examined to exclude regressed tumours.

400

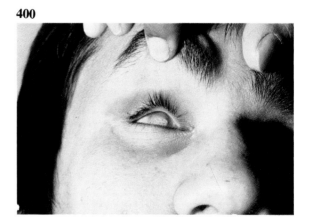

400   Norrie disease.
Note: Shrunken, opaque cornea and bulb ('pthisis bulbi').

Other features: Mental retardation in some, cataracts, retrolental opacities and sensorineural deafness.

Inheritance: X-linked recessive.

11 Skin disorders

401

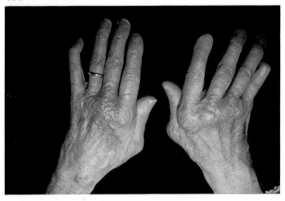

402

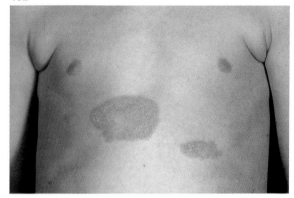

401 and 402   Psoriasis.
Note: Raised erythematous scaly patches.
  Pitting of nails, arthritis and pustules.
  Inheritance: Uncertain. The non-pustular form (psoriasis vulgaris) is associated with the HLA–A13 and HLA–A7 antigens.

403

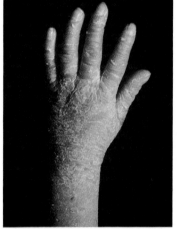

403   Ichthyosis vulgaris.
Note: Scaly, dry and erythematous skin.
  Other features: Flexural surfaces spared. Onset in childhood.
  Inheritance: Autosomal dominant. An X-linked form exists which may be distinguished clinically by involvement of flexural surfaces with onset in infancy in males. Affected males with the X-linked form have deficiency of steroid sulphatase.

404

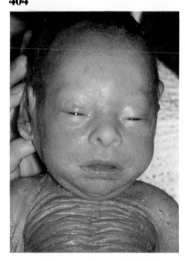

404   Lamellar ichthyosis.
Note: Collodion-like skin.
  Other features: Abnormal skin usually peels off to give good recovery.
  Inheritance: Autosomal recessive.

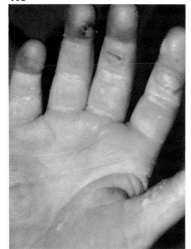 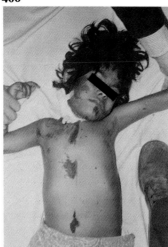 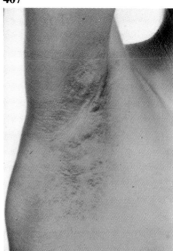

405   Palmo-plantar hyperkeratosis.
Note: Hyperkeratotic, peeling skin on palms.
  Other features: This can be associated with other features such as corneal dystrophy, periodontoclasia (Papillon–Lefevre), oesophageal cancer and diseases of other systems.
  Inheritance: Mostly autosomal dominant, those cases associated with corneal dystrophy or periodontoclasis may be autosomal recessive.

406   Linear sebaceous nevus.
Note: Pigmented, hyperkeratotic nevi affecting face and trunk. Striking demarcation at the midline.
  Other features: Seizures, mental retardation, cloudy corneae and conjunctival lipodermoids.
  Inheritance: Sporadic.

407   Acanthosis nigricans.
Note: Velvety, pigmented and thickened skin in the axilla. Papillomatous lesions are also present.
  Other features: Alopecia, palmoplantar hyperkeratosis, brittle nails and ocular involvement are sometimes seen; occasionally secondary to gastrointestinal carcinomata.
  Inheritance: Autosomal dominant.

408

 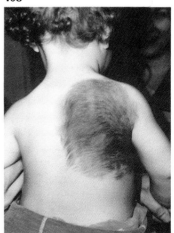

408   Giant pigmented hairy nevus.
Note: Large, pigmented and hairy nevus on back.
  Other features: Risk of melanoma, occasional intracranial abnormalities.
  Inheritance: The majority of cases are sporadic.

409

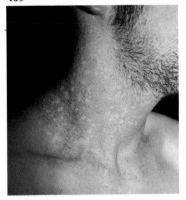

411

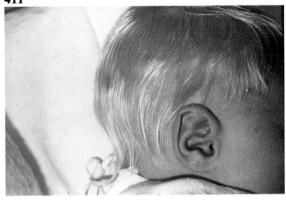

411 Oculo-cutaneous albinism.
Note: Absent pigment in skin and hair in affected children, red translucent irides (see also page 13).

Other features: Ocular problems (photophobia, nystagmus, decreased visual acuity), increased frequency of skin tumours (e.g. squamous cell carcinoma).

Inheritance: There are several genetic forms (see **Table 6**).

410

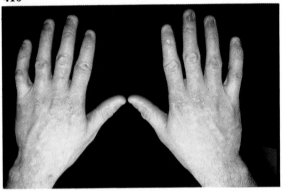

409 and 410 Dyskeratosis congenita.
Note: Hyperpigmentation of neck with hypo-pigmented areas giving a reticular pattern ('rain-drop pigmentation').

Similar skin lesions on hands and nail dystrophy.
Other features: Precarcinomatous leukoplakia of mucous membranes, pancytopenia.
Inheritance: X-linked recessive in most families.

412 Hypomelanosis of Ito.
Note: Irregular, depigmented patches.
Other features: Mental retardation, seizures and eye anomalies.
Inheritance: Autosomal dominant families have been described, although most cases are sporadic.

Table 6 Types of albinism

Type	Other features	Inheritance
Tyrosinase negative oculo-cutaneous	—	A.R.
Tyrosinase positive oculo-cutaneous	—	A.R.
Yellow mutant	—	A.R.
Ocular	—	X.L.R.
Hermansky–Pudlak syndrome	Bleeding diathesis	A.R.
Chediak–Higashi syndrome	Neutropenia, increased infections	A.R.
Cross syndrome	Ocular anomalies, spasticity and retardation	A.R.

412

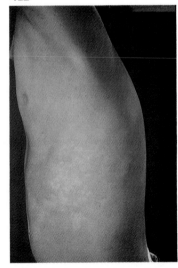

413

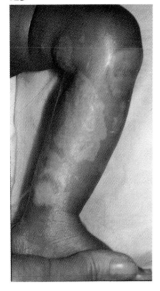

413 Epidermolysis bullosa simplex.
Note: Areas of hypopigmentation where bullae have ruptured.
Inheritance: Autosomal dominant.

414 Epidermolysis bullosa lethalis.
Note: Erosions and bullae on extremities and trunk.
Other features: Absent teeth and deformed nails.
Inheritance: Autosomal recessive.

415 Epidermolysis bullosa dystrophica.
Note: Scarring with fusion of the fingers, atrophic skin.
Other features: Bullae usually appear at birth, teeth and hair normal in the dominant form and thick dystrophic nails.
Inheritance: Autosomal dominant and recessive forms.

416 and 417 Incontinentia pigmenti.
Note: Multiple small bullae in a neonate, whorled skin pigmentation in an older child.
Other features: (In some cases) microcephaly, mental retardation, seizures, microphthalmia and other ocular anomalies, missing teeth and partial alopecia.
Inheritance: X-linked dominant, usually only females are affected. This is thought to be due to early death *in utero* of affected males.

414

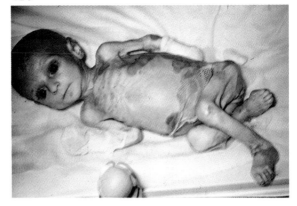

415

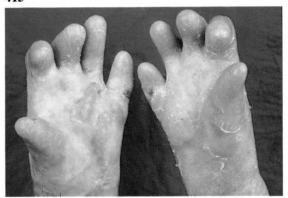

416

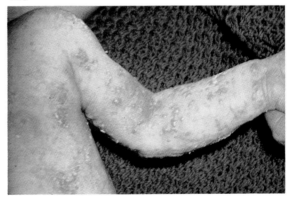

417

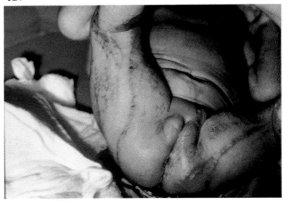

418

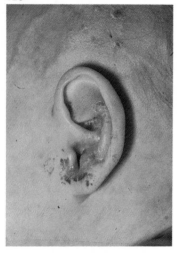

419

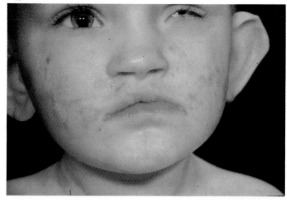

420

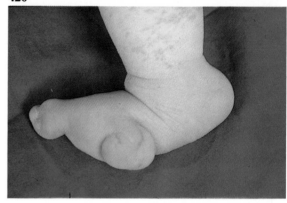

418   Acrodermatitis enteropathica.
Note: Vesicular, erythematous eruption of pinna.
Hair loss with associated pustules.
   Other features: Protracted diarrhoea, anorexia
and apathy. Symptoms respond to zinc treatment.
   Inheritance: Autosomal recessive.

419 and 420   Goltz syndrome.
Note: Microphthalmia, coloboma of iris, telangiectatic lesions and angiofibrotic nodules around
mouth.
   Syndactyly of toes, atrophy of skin.
   Other features: Hypoplastic teeth, dystrophic
nails, mental retardation, seizures, short stature,
scoliosis and congenital heart defect.
   Inheritance: X-linked dominant, thought to be
lethal prenatally in males.

422

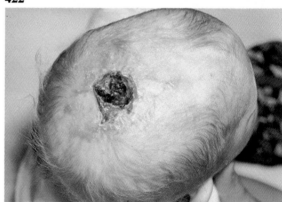

421

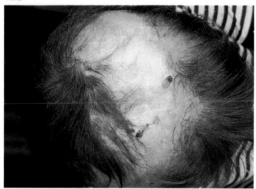

423

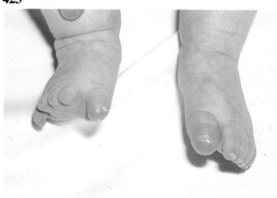

421–423   Aplasia cutis congenita.
Note: Alopecia and areas of atrophic skin with
scarring.
Other features: Can be associated with constriction rings of the limbs and ectrodactyly (see **423**).
   Inheritance: Autosomal dominant and recessive
forms exist. Scalp defects can be seen with trisomy
13 and the 4p- (Wolf–Hirschhorn) syndromes.

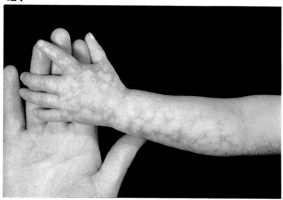

424

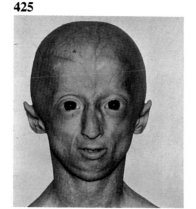

425

424 Poikiloderma congenita (Rothmund–Thomson).
Note: Erythroderma, scarring, depigmentation and telangiectasia giving 'marbled' appearance to skin.
Other features: Involvement of skin of face in a 'butterfly' distribution, cataracts, short stature, small hands with hypoplastic or absent thumbs, dystrophic nails, sparse hair and hypogenitalism.
Inheritance: Autosomal recessive.

425 Progeria.
Note: Early onset baldness, loss of subcutaneous fat and atrophic skin, premature ageing, thin, beaked nose, frontal bossing and small chin.
Other features: Short stature, hypoplastic nails, fibrosis around the joints, early onset of coronary artery disease and normal intelligence.
Inheritance: Uncertain, affected sibs have been described.

426

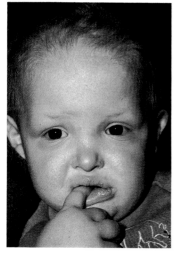

427

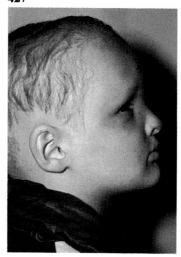

428

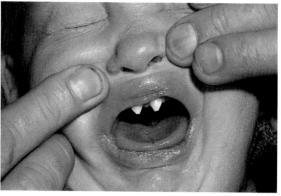

426–428 Anhydrotic ectodermal dysplasia.
Note: Hypotrichosis, thick lips, prominent frontal bossing, saddle-shaped nose and wrinkled skin around the eyes.
Absence of teeth or tendency to conical shape.
Other features: Reduced sweating and hyperthermia.
Inheritance: Mainly X-linked recessive with some manifestations in carrier females (mainly dental anomalies). Autosomal recessive families have also been described.

429

430

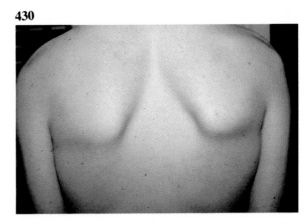

431

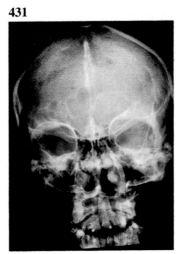

432

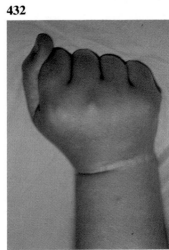

429 Parry–Romberg syndrome (progressive hemifacial atrophy).
Note: Unilateral atrophy of skin, soft tissue and underlying bone.
  Other features: Unilateral cerebral atrophy and Jacksonian epilepsy.
  Inheritance: Mainly sporadic.

430–432 Basal cell nevus syndrome.
Note: Multiple small nevi of skin, short meta-carpals and calcified falx cerebri.
  Other features: Jaw cyst, frontal bossing and large head, mild mental retardation, pits on palms and soles. Malignant change of skin nevi.
  Inheritance: Autosomal dominant.

433 Xeroderma pigmentosum.
Note: Skin freckling with white atrophic areas, alopecia and blepharitis.
  Other features: Skin tumours, the association with mental retardation, hypogonadism and microcephaly is known as the de Sanctis–Cacchione syndrome.
  Inheritance: Autosomal recessive. Biochemical defects of DNA-excision repair mechanisms have been described. There are at least five different biochemical complementation groups.

433

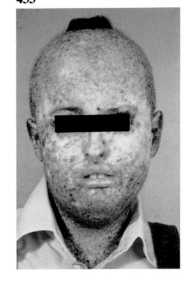

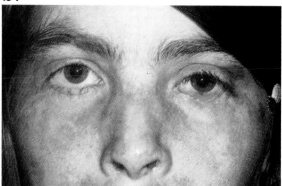

434 Bloom syndrome.
Note: Malar hypoplasia, telangiectatic erythema.

Other features: Short stature, microcephaly, *café au lait* skin pigmentation, immunoglobulin deficiency, increased number of *in vitro* chromosome breaks, increased sister chromatid exchange and propensity to develop lymphoreticular malignancy.

Inheritance: Autosomal recessive.

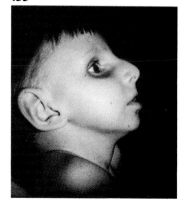

435 Cockayne syndrome.
Note: Microcephaly, slender nose and sunken eyes.

Other features: Growth deficiency, mental retardation, pigmentary retinopathy and photosensitive skin rash.

Inheritance: Autosomal recessive.

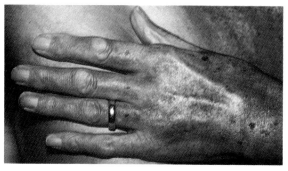

436 Variegate porphyria.
Note: Pigmentation of skin with atrophy and scarring.

Other features: Bullae, hypertrichosis and biochemical defects of porphyrin metabolism (see **Table 7**).

Inheritance: Autosomal dominant.

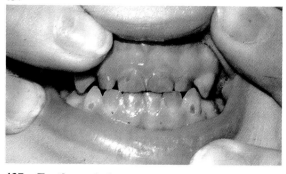

437 Erythropoietic porphyria.
Note: Reddish pigmentation of the teeth ('erythrodontia').

Other features: Splenomegaly, haemolytic anaemia, hypotrichosis, mutilation of digits and ears, and red urine.

Inheritance: Autosomal recessive (see **Table 7**).

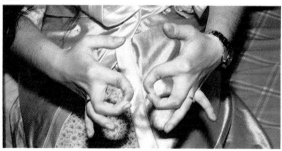

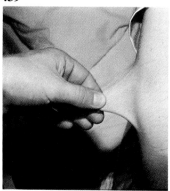

438 and 439 Ehlers–Danlos syndrome.
Note: Hyperextensible skin and joints.
Other features: See **Table 8**.
Inheritance: See **Table 8**.

Table 7 Classification of the porphyrias

Type	Clinical features	Inheritance	Urine	Faeces
Congenital erythropoietic (Günther disease)	Red urine, photo-sensitivity, nasal and aural cartilage erosion, mutilation of digits, erythrodontia and haemolytic anaemia	A.R.	Uroporphyrin I Coproporphyrin I	Coproporphyrin I ± Uroporphyrin I
Erythrohepatic protoporphyria	Photosensitivity, burning sensations	A.D.	Usually normal	Protoporphyrin
Acute intermittent porphyria	Acute abdominal attacks, peripheral neuropathy and neurological abnormalities. Acute attacks precipitated by alcohol and drugs (barbiturates)	A.D.	Porphobilinogen (PBG) d-aminolaevulinic acid (ALA)	Normal
Variegate porphyria	Photosensitivity, abdominal and neurological manifestations	A.D.	Copro-and uro-porphyrins PBG + ALA in acute attack	Porphyrins (copro + uro)
Hereditary coproporphyria	Similar to variegate porphyria	A.D.	Coproporphyrins	Coproporphyrins
Symptomatic porphyria	Photosensitivity	Acquired	Uroporphyrins	Coproporphyrins

Table 8 Inheritance of Ehlers–Danlos syndrome

Type	Clinical features	Inheritance
I (gravis)	Marked skin fragility, joint hyperextensibility. Bruising	A.D.
II (mitis)	As above, but less marked	A.D.
III (benign hypermobile)	Marked joint hyperextensibility. Hyperextensible but non-fragile skin	A.D.
IV (ecchymotic)	Rupture of blood vessels and intestines. Type III collagen deficiency	A.R.
V	Skin hyperextensibility and hypermobile finger joints. Lysyl oxidase deficiency	X.L.R.
VI (ocular)	Fragility of globes, severe scoliosis. Lysyl–hydroxylase deficiency	A.R.
VII (arthrochalasis multiplex congenita)	Multiple joint dislocations. Procollagen present	A.R.

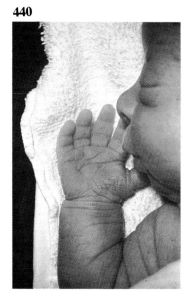 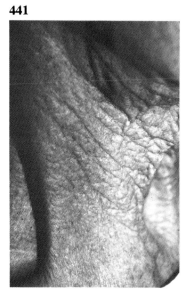 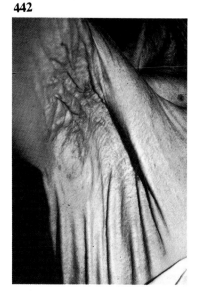

440   Cutis laxa.

Note: Sagging jowls, loose folds of skin and long upper lip.

Other features: Connective tissue elsewhere sometimes affected (gastro-intestinal, bladder, lungs and vocal cords).

Inheritance: X-linked recessive, autosomal recessive and autosomal dominant forms. Deficiency of lysyl oxidase has been noted in some X-linked forms.

441 and 442   Pseudo-xanthoma elasticum.

Note: Redundant folds and yellowish plaques at the neck and in the axilla.

Other features: Angioid streaks in retina, coronary and general arterial insufficiency and gastro-intestinal bleeding.

Inheritance: Autosomal dominant and recessive forms.

443

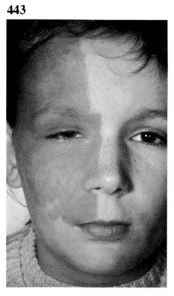
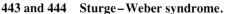

444

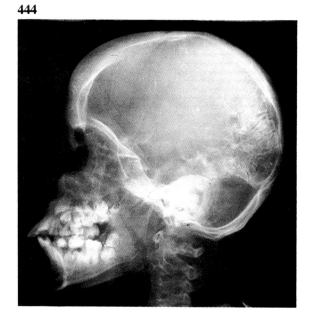

443 and 444   Sturge–Weber syndrome.

Note: Port wine stain involving divisions I and II of the trigeminal distribution, intracranial calcification.

Other features: Seizures, mental retardation, contralateral hemiplegia and buphthalmos.

Inheritance: Sporadic, recurrence risk low.

445

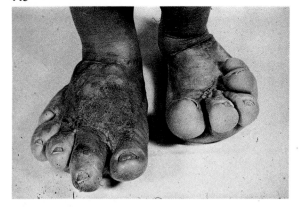

448

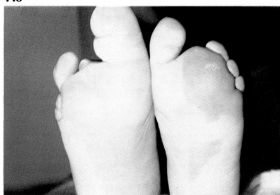

449

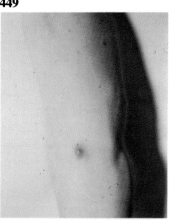

451

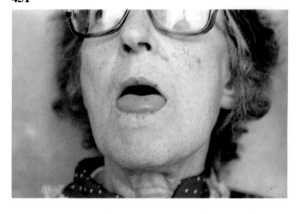

446 **447**

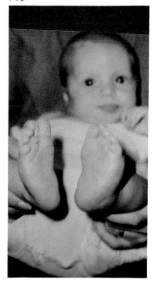 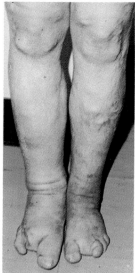

445–448   Klippel–Trenaunay–Weber syndrome.
Note: Asymmetrical hypertrophy of toes and feet, with associated vascular naevi.

Other features: Venous malformations of internal organs, syndactyly or polydactyly. Vascular lesions of the skin may be remote from the hypertrophied limb.

Inheritance: Usually sporadic.

449   Blue rubber bleb syndrome.
Note: Punctate blue naevi on arm.

Other features: Severe gastro-intestinal bleeding.

Inheritance: Mostly sporadic. Some autosomal dominant pedigrees described.

450

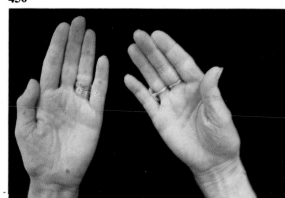

450 and 451   Osler–Rendu–Weber syndrome.
Note: Punctiform telangiectasia of skin and mucous membranes.

Other features: Gastro-intestinal bleeding, haemorrhages elsewhere (e.g. CNS, lungs, retina and urinary tract).

Inheritance: Autosomal dominant.

12 Neurological disorders

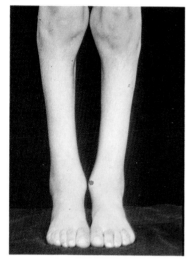

452

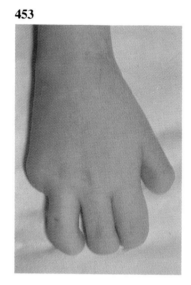

453

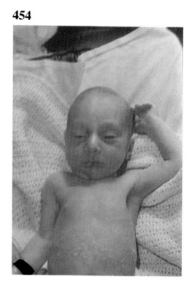

454

452 Hereditary motor and sensory neuropathy (Charcot–Marie–Tooth disease).
Note: Generalized muscle wasting, more pronounced peripherally in the lower limbs and bilateral pes cavus.

Inheritance: Type I (i.e. with reduced motor conduction velocities) is autosomal dominant with occasional recessive pedigrees described.

Type II (normal motor conduction velocities). Predominantly autosomal dominant, occasionally recessive.

453 Congenital insensitivity to pain.
Note: Mutilated hand due to reduction of pain sensation in a child with Type IV sensory neuropathy.

Inheritance: See **Table 9**.

454 Riley–Day syndrome (familial dysautonomia).
Note: Expressionless face with thin lips and dry skin in a child of Ashkenazi Jewish parents.

Other features: Feeding difficulties, hyperhydrosis, insensitivity to pain, lack of tears, smooth tongue, unstable temperature, short stature and retardation.

Inheritance: See **Table 9**.

Table 9 Sensory neuropathies

Type	Inheritance
I Hereditary sensory neuropathy	A.D.
II Congenital sensory neuropathy	A.R.
III Familial dysautonomia (Riley–Day)	A.R.
IV Congenital insensitivity to pain	A.R.
V Insensitivity to pain with anhydrosis	A.R.

455

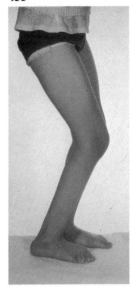

456

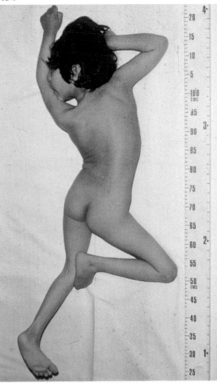

457

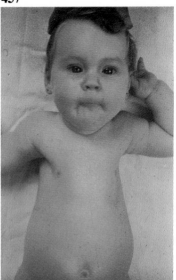

455 Spastic paraplegia.
Note: Flexion contractures of spastic limbs, pes planus.
 Other features: Spasticity of upper limbs, slow progression.
 Inheritance: See **Table 10**.

456 Cerebral palsy.
Note: Asymmetrical spastic quadriplegia.
 Other features and inheritance: See **Table 11** for types of cerebral palsy and recurrence risk.

Cerebral palsy – recurrence risks
A careful family and perinatal history must be taken in order to exclude definite genetic syndromes or cases due to perinatal injury. Recurrence risk for offspring of isolated cases are given in **Table 11**.

458

457 Infantile spinal muscular atrophy (Werdnig–Hoffman).
Note: Costal recession due to atrophy of respiratory muscles in a child with Werdnig–Hoffman disease.
 Other features: (See also **Table 12**), bulbar weakness, fasciculations and proximal limb muscle wasting.
 Inheritance: See **Table 12**.

458 Childhood onset spinal muscular atrophy (Kugelberg–Welander).
Note: Wasting of thenar, hypothenar, interosseous and other small hand muscles and scoliosis in a chair bound individual.
 Other features: Proximal wasting of all muscle groups, fasciculations and absent reflexes.
 Inheritance: Mostly autosomal recessive (see **Table 12** for risks to sibs of a sporadic case).

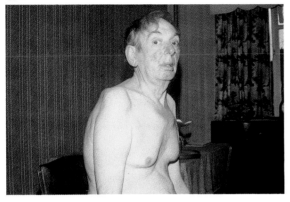 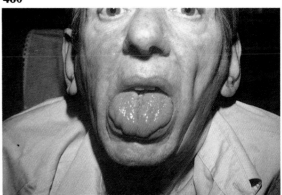

459 and 460   X-linked spinal muscular atrophy.
Note: Wasting of tongue and facial muscles.
   Proximal muscle wasting, gynaecomastia.
   Other features: Facial tremor.
   Inheritance: See **Table 12**.

Table 10   Genetics of spastic paraplegia

Familial types	
Autosomal dominant	70%
Autosomal recessive	30%

Table 11

Type	Risk to sibs
Hemiplegia	1 in 100
Diplegia	1 in 100
Ataxia	1 in 10
Dyskinetic	1 in 10
Symmetric spastic paraplegia	1 in 10
Double hemiplegia	
Asymmetric spastic quadriplegia	1 in 100
Mixed cerebral palsy	

Table 12   Classification of proximal spinal muscular atrophies

Type	Onset	Course	Risk to sibs of sporadic case	Risk to offspring
Werdnig–Hoffman	Before 6 months	Death before 3 years	1 in 4	—
Childhood onset	Before 3 years	Chronic	1 in 5	1 in 50
Kugelberg–Welander	Between 3 and 18 years	Chronic	1 in 10	1 in 10
Adult onset	3rd decade onwards	Chronic	?	1 in 20
X-linked type	3rd decade	Chronic	X-linked recessive inheritance	

461

462

463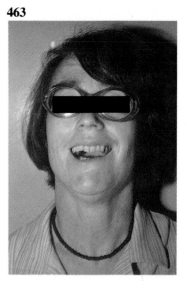

464

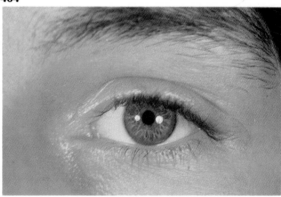

465

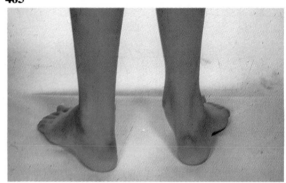

461 Motor neurone disease.
Note: Weakness and wasting of the small muscles of the hand.
Other features: Bulbar and pyramidal tract involvement.
Inheritance: About 10% are genetic (autosomal dominant). Risk to sibs and offspring of isolated cases, small.

462 Torsion dystonia.
Note: Fixed posturing of hands and arms.
Other features: Torticollis, severe functional disability.
Inheritance: There are two types:
(a) Autosomal recessive. This is more common in Ashkenazi Jews. Onset tends to be before 20 years of age.
(b) Autosomal dominant. Onset tends to be after 20 years of age.

463 Huntington's chorea.
Note: Facial grimacing.
Other features: Choreiform movements, dementia and brisk reflexes. Average age of onset 35–45 years of age.
Inheritance: Autosomal dominant.

464 Wilson's disease.
Note: Kayser–Fleischer rings at periphery of iris showing as a rim of brown pigment.
Other features: Onset in childhood, liver disease, dementia and choreiform movements with athetosis.
Inheritance: Autosomal recessive.

465 Friedreich's ataxia.
Note: Pes cavus.
Other features: Onset before 20 years, ataxia with dysarthria, extensor plantar responses, absent reflexes, scoliosis, heart defects, posterior column sensory loss and diabetes.
Inheritance: Autosomal recessive. This condition must be differentiated from adult onset cerebellar ataxia with ophthalmoplegia, dementia and normal or brisk reflexes, which is usually autosomal dominant.

466 Ataxia-telangiectasia.
Note: Conjunctival telangiectasia.
Other features: Progressive ataxia, dementia, frequent infections, high incidence of lymphoid tumours, immune deficiency, chromosome breakage and raised serum AFP.
Inheritance: Autosomal recessive.

466

467–469 Microcephaly-autosomal recessive type.
Note: Small head circumference with receding forehead and normal sized face. The ears appear to be relatively large.
Other features: Mental retardation, minimal spasticity and absence of seizures.
Inheritance: Environmental causes such as intra-uterine infections must be excluded. The diagnosis of the autosomal recessive type is based on characteristic head shape, absence of environmental causes and presence of parental consanguinity or affected sibs. The risk to sibs of an isolated case is 1 in 8.

467

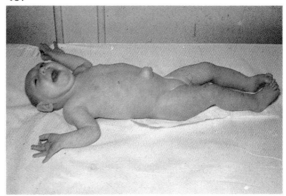

468

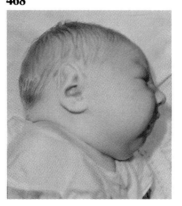

469

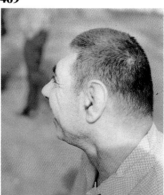

470

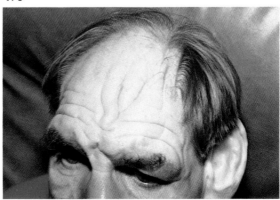

470 Cutis verticis gyrata.
Note: Microcephaly, furrowing and folding of scalp.
Other features: Mental retardation, seizures and eye defects.
Inheritance: Possibly autosomal recessive.

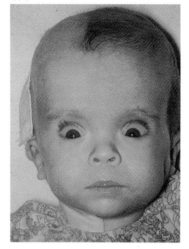

471   Hydrocephalus.

Note: Large cranium, prominent scalp veins, 'sunset sign' to eyes and recent shunt operation.

Other features: Neurological dysfunction if untreated.

Inheritance: In general only the type due to aqueduct stenosis in males carries a high risk (there is an X-linked recessive form). The risk for male sibs of an isolated case of aqueduct stenosis is 1 in 10. For all other causes the recurrence risk is about 1–2%, unless there is a positive family history.

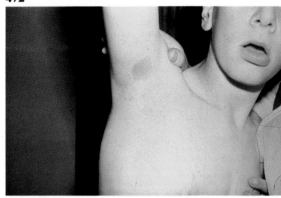

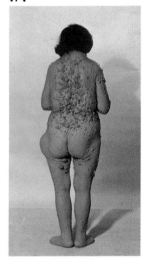

472–474   Neurofibromatosis (Von Recklinghausen).

Note:

(a) *Café au lait* spot and freckles in axilla.

(b) Multiple neurofibromata on limbs and trunk.

(c) Severe, multiple neuromata.

Other features: Pseudarthrosis of the tibia, scoliosis, neoplasms including meningiomas, gliomas, phaeochromocytomas, neurofibrosarcoma, acoustic neuromas, mental retardation (5–10%) and hypertension.

Inheritance: Autosomal dominant. Fifty per cent of cases are fresh mutants. High penetrance but variable expression.

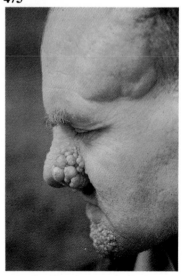

475–479   Tuberose sclerosis.

Note:

(a) Fibrous angiomatous lesions in characteristic distribution at angle of nose and chin, but sparing the upper lip and Shagreen patch on forehead.

(b) Less severe lesions ('adenoma sebaceum').

(c) Amelanotic macule.

(d) Subungual fibromata.

(e) Intracranial calcification in peri-ventricular distribution.

Other features: Retinal phakomata, epilepsy, mental retardation, rhabdmyosarcomata, renal tumours and cerebral tumours.

Inheritance: Autosomal dominant with variable expression. Seventy-five per cent of cases are fresh mutants. Persons at risk for carrying the abnormal gene can be screened by looking at the skin with Wood's lamp and by carrying out a CAT scan.

476

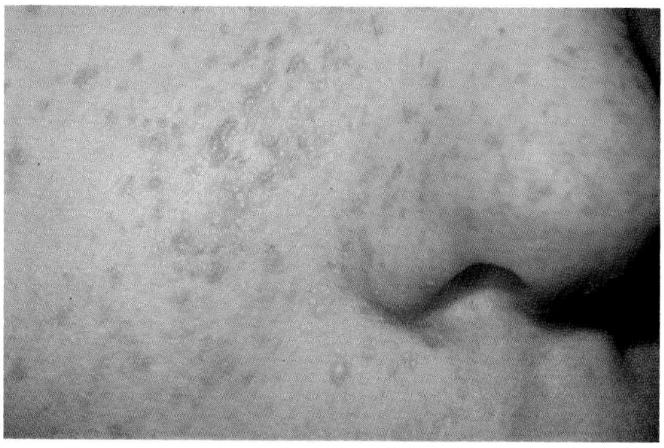

477

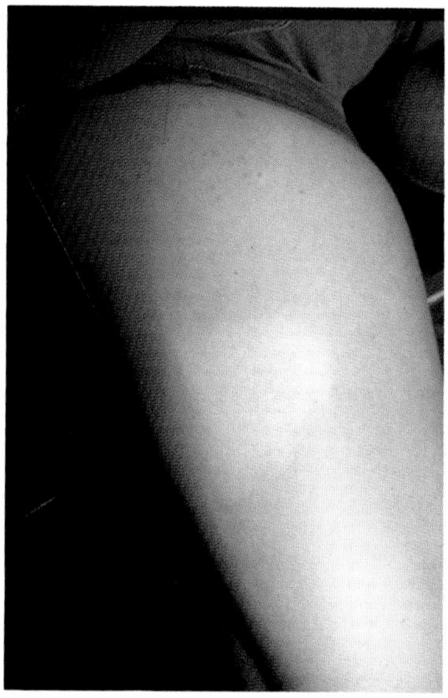

478

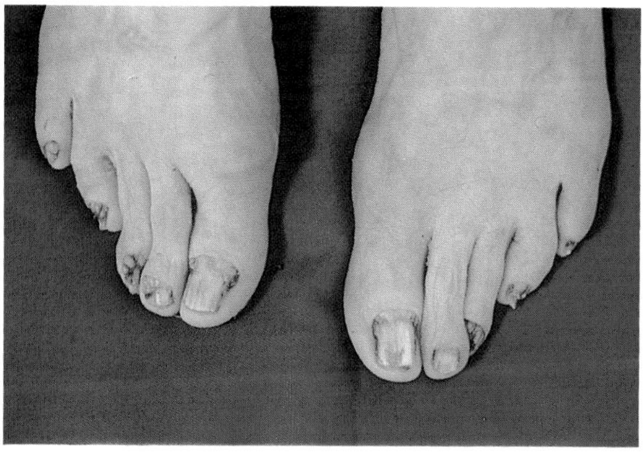

479

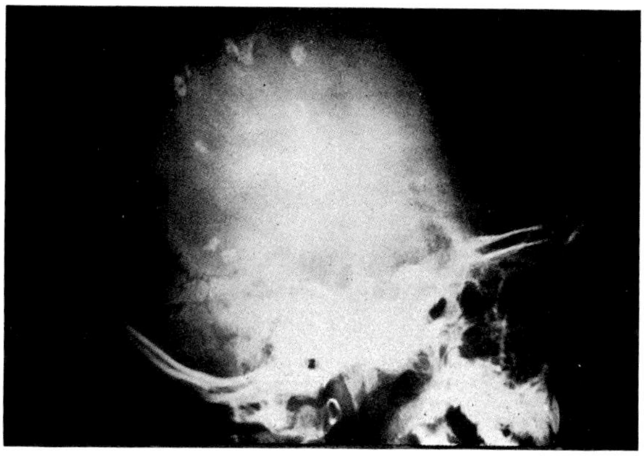

480

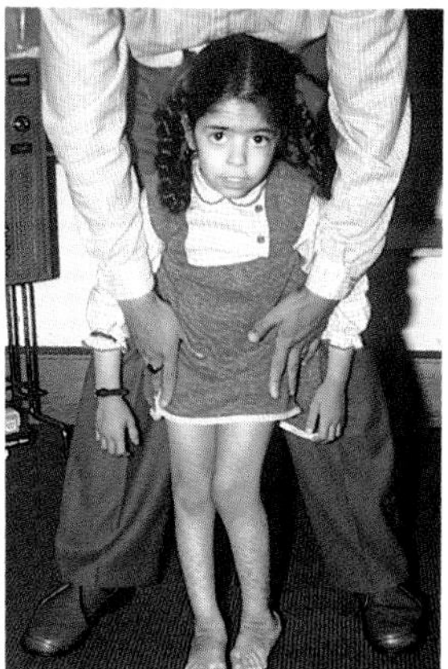

481

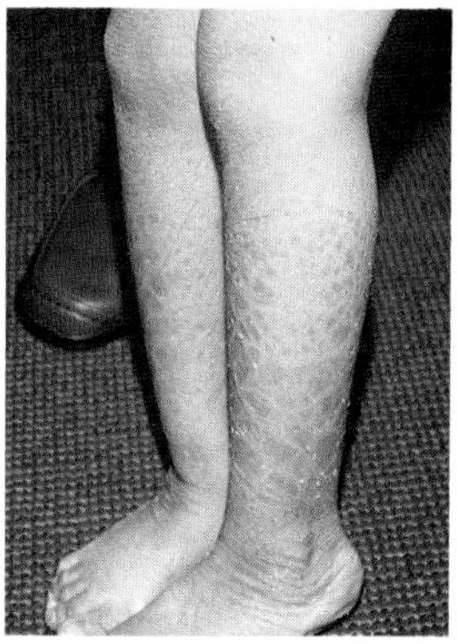

482

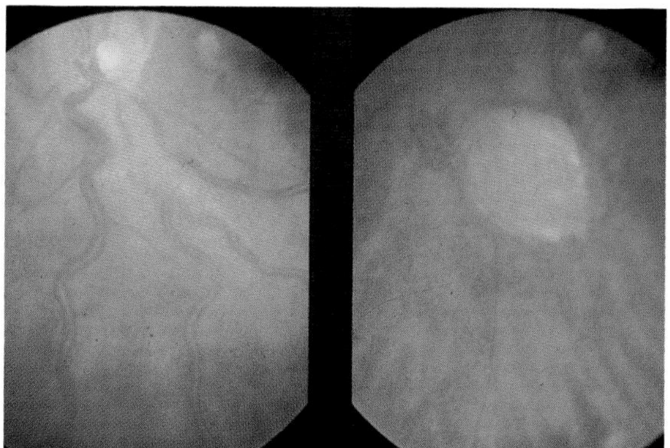

480 and 481 Sjögren–Larsson syndrome.
Note: Ichthyosis of lower limbs, spasticity of lower limbs.

Other features: Mental retardation, atypical retinitis pigmentosa, 'glistening' lesions of the macula.

Inheritance: Autosomal recessive.

482 Von Hippel–Lindau syndrome.
Note: Retinal angiomata, tortuous and dilated vessels.

Other features: Cerebellar haemangioblastoma, cystic lesions of kidney, renal tumours and phaeochromocytomas.

Inheritance: Autosomal dominant.

13 Muscle disorders

483 Duchenne muscular dystrophy.
Note: Pseudohypertrophy of calves, proximal wasting in boy. Sister is a possible carrier.

Other features: Onset 2–3 years, mental retardation in some, scoliosis, cardiomyopathy, wheelchair bound by 11 years and death by end of 2nd decade.

Inheritance: X-linked recessive. One-third of all cases are thought to be new mutants. Measurement of serum creatine kinase in possible female heterozygotes may be used for carrier detection. Results of three separate tests should be combined statistically with pedigree data in order to arrive at a final probability.

483

484

484 Becker dystrophy.
Note: Pseudohypertrophy in a teenage boy who is still ambulent.

Other features: Proximal muscles of shoulder and pelvic girdle affected. Onset in 1st or 2nd decade, still ambulant at 11 years, generally not wheelchair bound before 3rd decade and occasional cardiomyopathy.

Inheritance: X-linked recessive.

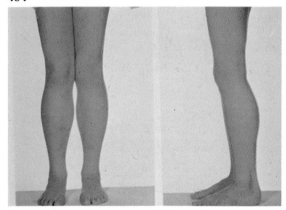

485

485 Limb-girdle dystrophy.
Note: Wasted pelvic and shoulder girdle muscles, with relative sparing of the deltoids.

Other features: Onset in 2nd to 3rd decade, pseudohypertrophy of calves in a small proportion. Must be differentiated from the Becker type in a male.

Inheritance: Autosomal recessive.

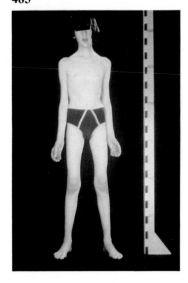

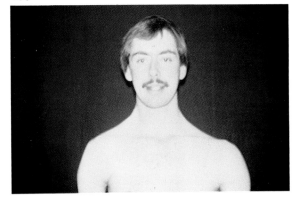

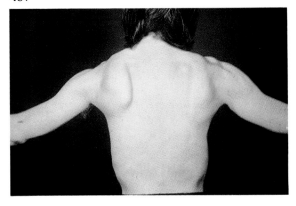

488

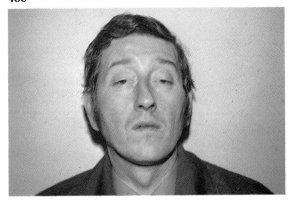

486 and 487   Facio-scapulo-humeral dystrophy.
Note: Shoulder girdle atrophy with winging of the scapulae, facial weakness.
Other features: Onset variable from childhood to early adulthood. Features can be very mild.
Inheritance: Autosomal dominant.

488   Ocular myopathy.
Note: Ptosis.
Other features: Progressive external ophtho-myoplegia. See also **Table 13**.
Inheritance: See **Table 13**.

489   Congenital myopathy.
Note: Exaggerated lordosis and predominantly proximal myopathy.
Other features and inheritance: See **Table 14** for various types.

489

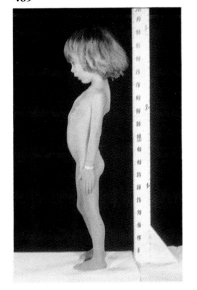

Table 13

Ocular myopathy: in its simplest form (bi-lateral ptosis), inheritance is as a dominant, but many cases progress to involve other muscles.
   Ptosis and ophthalmoplegia: inheritance is mostly dominant
   Ptosis, ophthalmoplegia and other muscles: inheritance mostly dominant
   Ocular pharyngeal muscular dystrophy: inheritance is usually dominant
   N.B. Mitochondrial myopathies should be excluded.

490

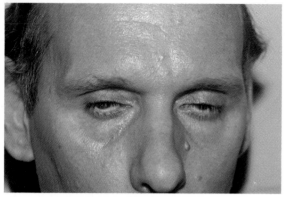

491

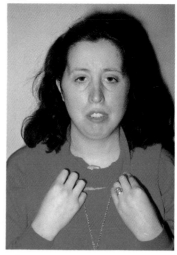

490 and 491   Myotonic dystrophy.
Note: Myopathic, 'snarling' smile, ptosis, frontal baldness, sternomastoid weakness and myotonic posturing of hands.

Other features: Cataracts, cardiomyopathy, peripheral muscle weakness. Clinical features can be very variable.

Inheritance: Autosomal dominant. The myotonic dystrophy locus is linked to the secretor locus. Cataracts predate other clinical features and must be looked for by slit-lamp examination in possible gene carriers. Children of mothers with this syndrome are at risk for severe congenital myotonic dystrophy.

492

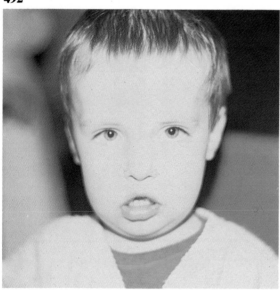

492   Congenital myotonic dystrophy.
Note: Weakness of muscles around eyes and mouth ('myopathic facies').

Other features: Mental retardation, swallowing difficulties and respiratory problems.

Inheritance: Autosomal dominant. Most infants with congenital myotonic dystrophy are the children of mothers with adult onset myotonic dystrophy.

Table 14
Modes of inheritance of congenital myopathies

Disease	Mode of inheritance
Nemaline myopathy	Both dominant and recessive pedigrees are known
Central core disease	Autosomal dominant
Myotubular or centronuclear myopathy	X-linked – severe (early onset)
	Dominant – mostly
	Recessive – rare
Fibre type disproportion	Probably autosomal dominant with variable expression
Congenital muscular dystrophy	Autosomal recessive
Mitochondrial myopathies	**Mode of inheritance**
Severe infantile	Mostly recessive
Milder with later onset, with or without lactic acidosis	Dominant or recessive
Kearns–Sayre	Mostly sporadic, occasional dominant and recessive pedigrees

14 Metabolic disorders

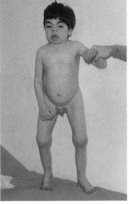 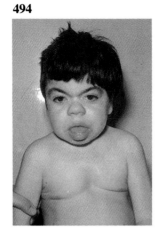

493 **494**

493 and 494 Hurler disease (mucopolysaccharidosis type I).

Note: 'Coarse' facial features consisting of prominent supra-orbital ridges, thick eyebrows, depressed nasal bridge, broad bulbous nose, thick lips and large, protruding tongue, 'snuffly' nose and coarse hair.

Prominent joints with flexion deformities.

Other features and inheritance: See **Table 15**.

495 **496**

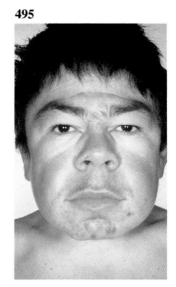 

497

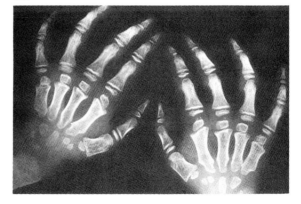

498

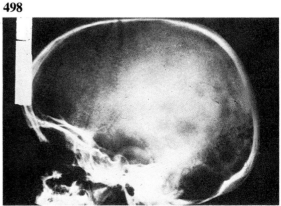

495–498 Hunter disease.

Note:

(a) 'Beaked' lumbar vertebrae with platyspondyly.

(b) Proximal tapering of metacarpals, coarse trabeculation and irregular epiphyses.

(c) Enlarged 'J'-shaped sella turcica.

(d) 'Coarse' facial features, prominent supraorbital ridges and hirsutism.

Other features and inheritance: See **Table 15**.

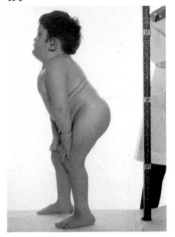

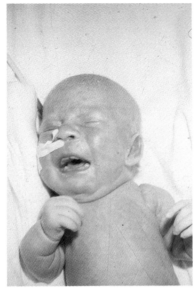

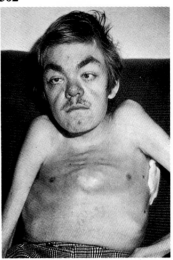

499 Morquio disease (mucopolysaccharidosis type IV).
Note: Short trunk, prominent joints, joint restriction, genu valgum and pectus carinatum.
   Other features and inheritance: See **Table 15**.

503

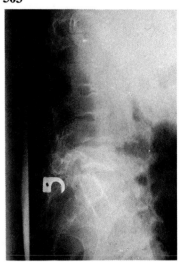

501

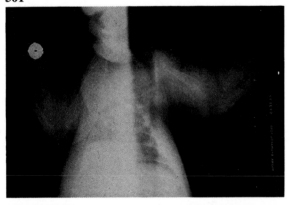

500 and 501   Mucolipidosis II (I-cell disease).
Note:
(a) Long prominent philtrum, gum hypertrophy and wide alveolar margins, smooth, tight skin and short neck.
(b) Periosteal cloaking of long bones.
   Other features: Progressive coarsening of facial features, growth deficiency, mental retardation and kyphosis.
   Inheritance: Autosomal recessive. Levels of multiple lysosomal hydrolases are raised in the serum. Antenatal diagnosis possible.

502 and 503   Mucolipidosis I (sialidosis type 2).
Note: Coarse 'Hurleroid' facial features, strabismus, pectus carinatum. Ossification defects of antero-superior parts of lumbar vertebrae.
   Other features: Myoclonus, progressive dementia, cherry-red spot at macula, deafness and hepato-splenomegaly occasionally. Deficiency of neuraminidase (and ß-galactosidase in some types). Abnormal oligosaccharides in the urine.
   Inheritance: Autosomal recessive. Sialidosis type 1 also has deficiency of neuraminidase with myoclonus and cherry-red spots at the macula, however there are no dysmorphic features.

Table 15   Mucopolysaccharidoses

Type	Inheritance	Age of onset	Cornea	Mental retardation	Skeletal changes	Urine MPS	Enzyme deficiency
IH (Hurler)	A.R.	1st year	Cloudy	++	+++	Dermatan + heparan sulphate	∝-L-iduronid-ase
IS (Scheie)	A.R.	5–10 years	Cloudy	−	+	Dermatan + heparan sulphate	∝-L-iduronid-ase
II (Hunter) (mild and severe forms)	X.L.R.	0–5 years	−	+/−	++	Dermatan + heparan sulphate	Iduronate sulphatase
III (Sanfilippo) (three types)	A.R.	1st decade	+/−	+++	+/−	Heparan sulphate	Heparan-N-sulphatase (type A)
IV (Morquio)	A.R.	2–5 years	+	−	+++	Keratan sulphate	Galactosamine-6-sulphate-sulphatase
VI (Maroteaux-Lamy)	A.R.	2–5 years	+	−	+++	Dermatan and chondroitin sulphate	Arylsulphatase B
VII (Sly)	A.R.	1st year	+/−	+	++	Dermatan and heparan sulphate	ß-Glucuronid-ase

504   Gaucher disease.
Note: Hepato-splenomegaly.

Other features: Gaucher disease is classified into three clinical types:
(a) Type A (infantile) is characterized by onset at 4–6 months of age with hypertonicity, opisthotonus, epilepsy, hepato-splenomegaly and death before one year of age.
(b) Type B (juvenile) is characterized by onset between 6 months and a year with progressive dementia, ataxia, eye movement disorders, epilepsy and hepato-splenomegaly
(c) Type C (adult) is characterized by an absence of neurological symptoms, but hepato-spleno-megaly and bony lesions are features.

All types have a deficiency of glucocerebrosidase. Antenatal diagnosis is possible.
Inheritance: Autosomal recessive.

505

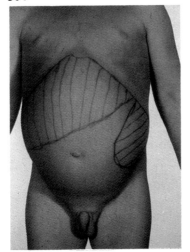

505   Neurovisceral storage disease with sea-blue histiocytes.
Note: Lipid-laden ('sea-blue') histiocyte.

Other features: Hepato-splenomegaly, progressive supranuclear ophthalmoplegia, intellectual decline. There is a possible relationship with Niemann–Pick disease type C. (For inheritance and clinical features of Niemann–Pick disease see **Table 16.**)
Inheritance: Autosomal recessive.

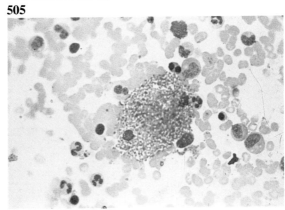

Table 16 Clinical types of Niemann–Pick disease

Type	Onset	Clinical features
A (acute neuropathic)	1st year	Progressive motor and mental deterioration, cherry-red spot at macula, hepato-splenomegaly, death by 3 years
B (chronic)	Childhood	Hepato-splenomegaly
C (juvenile or subacute)	1–6 years	Epilepsy, spasticity, intellectual deterioration, hepato-splenomegaly, death in late childhood
D (Nova-Scotia variant)	Childhood	Similar to type C – all patients have come from Nova-Scotia
E (adult)	Adulthood	No neurological abnormalities, hepato-splenomegaly

507

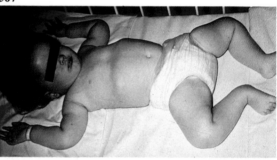

508

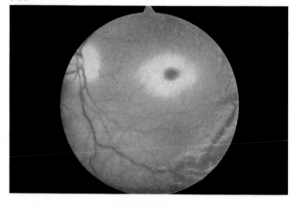

506

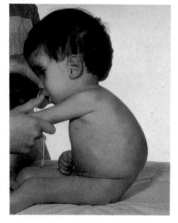

506   GM$_1$ gangliosidosis (severe infantile type).
Note: Coarse facial features, frontal bossing, low nasal bridge and thoracolumbar kyphosis.

Other features: Onset in infancy, 'cherry-red' spot at macula, hypotonia, growth and mental deficiency, hepatomegaly, vacuolated lymphocytes, seizures and X-rays show periosteal cloaking in newborn period with dysostosis multiplex appearance later. Deficiency of ß-galactosidase.

Inheritance: Autosomal recessive. Antenatal diagnosis possible. There are also other forms of ß-galactosidase deficiency with milder clinical features.

507–509   GM$_2$ gangliosidosis.
Note: Cherry-red spot at macula. Frog-like posture in severely floppy infant. Macrocephaly in an older child.

Other features: Onset at around 6 months of age with seizures, impaired vision and hypotonia. Death by 3 years.

Inheritance: Autosomal recessive. The syndrome is more common in infants of Ashkenazi Jewish extraction. Carrier detection is possible and to be encouraged in Ashkenazi Jews. Antenatal diagnosis possible.

Table 17

Type	Onset	Clinical features
Late infantile	1–2 years	Irritability, loss of ability to maintain posture, peripheral neuropathy and myoclonus
Juvenile	3–21 years	Subtle mental deterioration, peripheral neuropathy variable, athetoid posturing
Adult	over 21 years	
Multiple sulphatase deficiency (Austin variant)	1–2 years	'Hurleroid' facial features, hepato-splenomegaly, neurologically similar to late infantile type

510   Krabbe disease.
Note: Severe spasticity with opisthotonos, debility and wasting.

Other features: Onset within the first few months of life, fever and irritability, loss of tendon reflexes, myoclonus and seizures, death by 3 years, deficiency of galactocerebroside ß-galactosidase, antenatal diagnosis available.

Inheritance: Autosomal recessive.

511   Metachromatic leucodystrophy.
Note: Spasticity with decerebrate posture, peripheral wasting.

Other features: Onset (see **Table 17**), deficiency of arylsulphatase A, prenatal diagnosis available.

Inheritance: Autosomal recessive.

512   Ceroid lipofuscinosis.
Note: Retinal pigmentary degeneration, thin vessels and small yellow lesions.

Other features: For clinical features see **Table 18**. The presumed enzyme defect is unknown. Diagnosis can be made by rectal biopsy.

Inheritance: See **Table 18**.

510

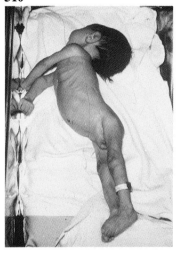

511

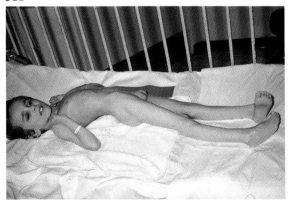

512

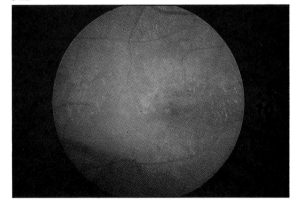

Table 18 Types of ceroid lipofuscinosis

Type	Onset	Clinical features	Inheritance
Infantile (Santavuori)	0–1 year	Ataxia, myoclonus, visual deterioration	A.R.
Late infantile (Jansky–Bielchowsky)	1–4 years	Myoclonus, convulsions, ataxia and visual loss	A.R.
Juvenile (Spielmeyer–Vogt)	5–10 years	Visual loss (pigmentary retinopathy) – seizures late	A.R.
Adult (Kufs)	Adult	Ataxia, myoclonus	A.D.

513

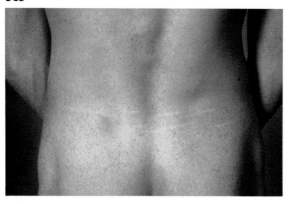

513 Fabry disease.
Note: Punctate angiomatous lesions of the skin.
Other features: Corneal opacities, mild coarseness of face, occasional mental retardation, spontaneous 'burning' pains in the limbs and seizures. Deficiency of ∝-galactosidase.
Inheritance: X-linked recessive.

514

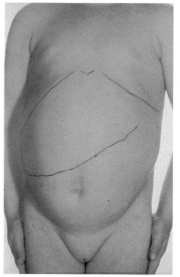

514 Glycogen storage disease type 1 (Von Gierke).
Note: Gross hepatomegaly.
Other features: Hypoglycaemia, cirrhosis and deficiency of glucose-6-phosphatase.
Inheritance: Autosomal recessive.

515

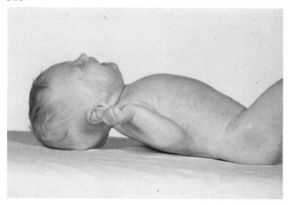

515 Phenylketonuria.
Note: Pale, sparse hair and eczematous skin eruption.
Other features: Seizures and mental retardation if untreated by dietary reduction of phenylalanine. Neonatal screening is now well established.
Inheritance: Autosomal recessive.

516

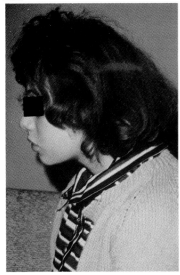

517

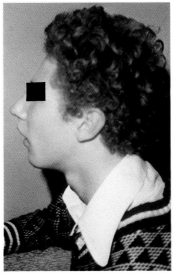

518

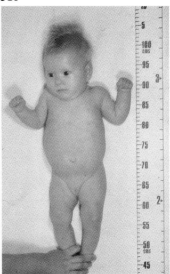

516 and 517   Phenylketonuria (maternal effects).
Note: Microcephaly in son and daughter of an untreated mother with phenylketonuria.

518   Homocystinuria.
Note: Fine, sparse hair.
Other features: Downward subluxation of the lens, malar flush, seizures, variable mental retardation, slim build, arachnodactyly, arterial and venous thromboses.
Inheritance: Autosomal recessive.

519

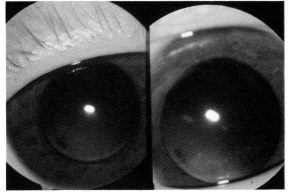

519   Galactosaemia.
Note: Early cataracts.
Other features: Jaundice, vomiting, failure to thrive, reducing substances in urine, fatal if untreated by diet. Deficiency of galactose-1-phosphate uridyl transferase.
Inheritance: Autosomal recessive.

520 and 521   Cystinosis.
Note: Fine corneal opacities, intracellular cystine crystals.
Other features: Fair hair, photophobia, aminoaciduria, polyuria and polydipsia, acidosis and hypokalaemia, hypophosphataemic rickets and short stature.
Inheritance: Autosomal recessive.

520

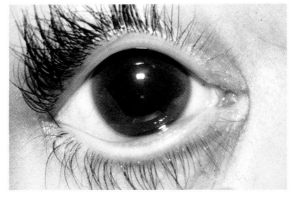

521

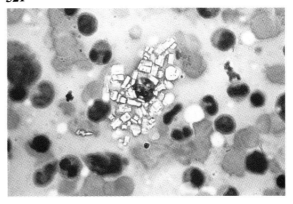

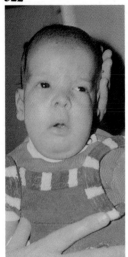
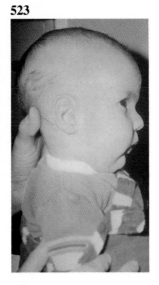

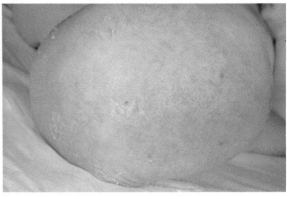

522 and 523   Zellweger syndrome.
Note: High forehead, expressionless face, micrognathia and shallow supra-orbital ridges.

Other features: Hypotonia, nystagmus, hepatomegaly, severe retardation, albumen and pipecolic acid in urine, high serum iron and iron binding capacity, stippled epiphyses, renal cysts, polymicrogyria and defects of myelination and death before 6 months.

Inheritance: Autosomal recessive.

524   Menkes syndrome.
Note: Sparse, stubby ('kinky') hair.

Other features: Progressive cerebral degeneration, seizures, tortuosity of arteries, low serum copper, coarse trabeculation and fragmentation of metaphyses and hairs are twisted and fragile when viewed under the microscope.

Inheritance: X-linked recessive. Female carrier detection and antenatal diagnosis possible.

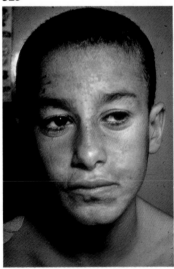

525   Lipoid proteinosis.
Note: Atrophic scars as the result of healing of impetiginous lesions. Nodules and general waxy appearance of skin.

Other features: Hoarseness of voice, nodules in other organs including tongue, vocal cords and buccal mucosa.

Inheritance: Autosomal recessive.

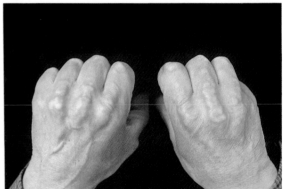

526   Hyperlipidaemia type II.
Note: Xanthomata on dorsum of hands and fingers.

Other features: Tendinous xanthomata, ocular xanthelasma and arcus senilis. Early arteriosclerosis. Individuals with this disorder have abnormalities of cell surface lipoprotein receptors.

Inheritance: Autosomal dominant.

527 Berardinelli lipodystrophy syndrome.
Note: Atrophy of subcutaneous fat giving the appearance of hypertrophied muscles, hyperpigmentation around the neck.

Other features: Hyperlipidaemia, hirsutism, accelerated growth and maturation, large hands and feet, enlargement of the genitalia and occasional mental retardation.

Inheritance: Autosomal recessive.

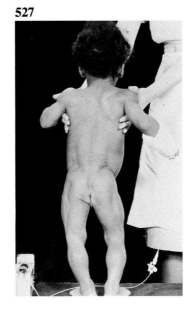

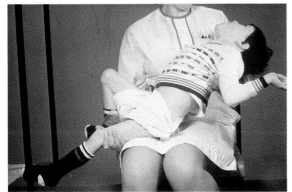

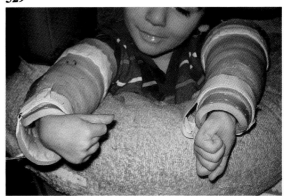

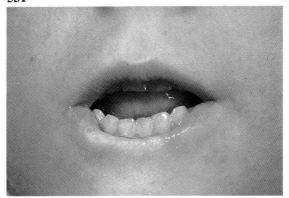

528–531 Lesch–Nyhan syndrome.
Note: Dystonic posturing, self-mutilation of lips and splinted arms to prevent self-mutilation.

Other features: Apperent dementia, high plasma uric acid, deficiency of hypoxanthine – guanine – phosphoribosyl – transferase (HGPRT).

Inheritance: X-linked recessive, carrier detection and antenatal diagnosis possible.

15 Genito-urinary disorders

532

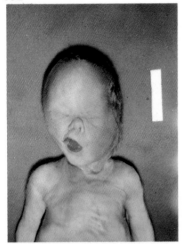

533

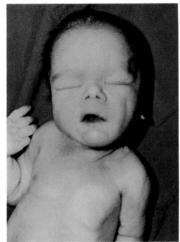

534

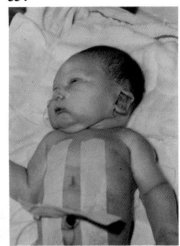

535

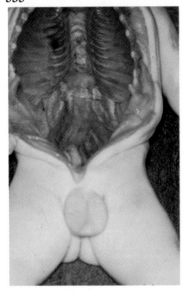

532–535 Renal agenesis.
Note:
(a) Bilateral absence of kidneys.
(b) Compressed facial features, infra-orbital creases.

Other features: Talipes equinovarus, pulmonary hypoplasia.

Inheritance: Recurrence risk for sibs of an isolated case, 3%. If either parent has unilateral renal agenesis, the risk is around 10%. Antenatal diagnosis possible by ultrasound.

536

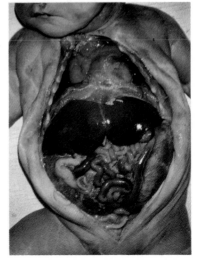

536 and 537 Polycystic kidneys (infantile form).
Note: Grossly enlarged kidneys, radial cysts of the kidney.

Other features: Multiple cysts in kidneys and liver. Early demise.

Inheritance: Autosomal recessive. This condition must be differentiated from (a) cystic dysplastic kidneys, (b) adult autosomal dominant polycystic kidney disease with early onset, by detailed histological analysis.

537

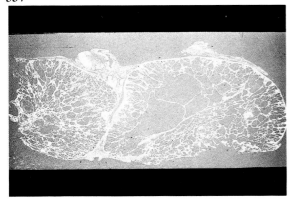

338

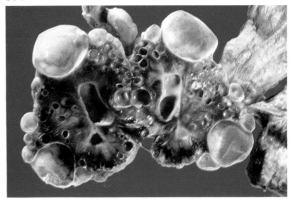

538 Cystic dysplastic kidney.
Note: Grossly disorganized cystic kidney secondary to ureteric obstruction.

Other features: Oligohydramnios and Potter facies if the obstruction is bilateral.

Inheritance: Usually sporadic. Recurrence risk low.

539

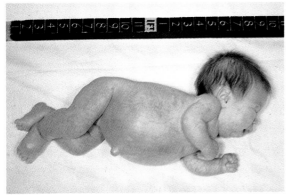

540

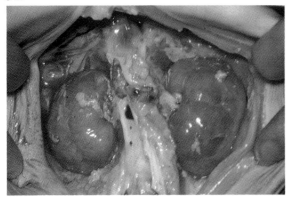

539 and 540 Congenital nephrosis.
Note:
(a) Bilaterally enlarged kidneys.
(b) Gross oedema and distension of the abdomen.

Other features: Onset at birth of proteinuria and oedema. Early death.

Inheritance: The so-called Finnish type is autosomal recessive, although there may be heterogeneity in the group as a whole. Antenatal diagnosis is possible through raised amniotic fluid AFP.

541

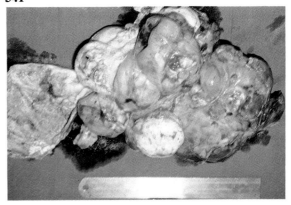

541 Wilm tumour.
Note: Extensive tumour in an excised kidney.

Other features: Wilm tumour can be associated with hemihypertrophy or aniridia.

Inheritance: There is evidence for a strong genetic component in Wilm tumour, especially in bilateral cases. The recurrence risks to sibs of a bilateral case are around 10%. For a unilateral case they are 5%. Small deletions of the short arm of chromosome 11 can produce the so-called AGR triad of aniridia, Wilm tumour, genital anomalies and mental retardation.

542

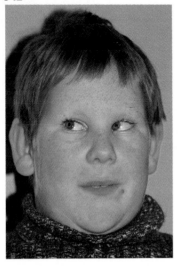

543

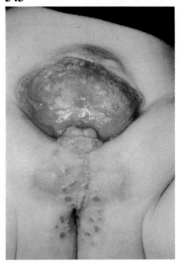

544

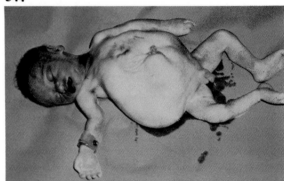

545

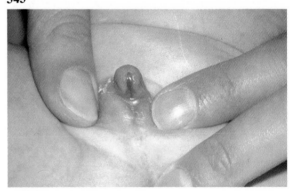

546

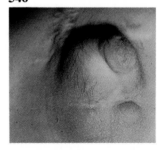

542   Lowe syndrome.
Note: Microphthalmia, short stature and prominent forehead.
Other features: Mental retardation, cataracts and renal tubular dysfunction with amino-aciduria.
Inheritance: X-linked recessive.

543   Exstrophy of the bladder.
Note: Exposed bladder mucosa. Genital anomalies.
Inheritance: Mostly sporadic, recurrence risk less than 1%.

544   Prune belly syndrome.
Note: Grossly distended abdomen with deficient abdominal musculature and 'Potter's' facies.
Other features: Urethral or bladder neck obstruction. Mega-ureter and hydronephrosis. High male to female sex ratio.
Inheritance: Usually sporadic. Recurrence risk for sibs is less than 1%.

545   Hypospadias.
Note: Urethral opening on ventral aspect of the penis. Deficiency of foreskin.
Other features: This can be a feature of many dysmorphic syndromes, so that associated malformations must be looked for.
Inheritance: Recurrence risk for a male sib of an isolated case, about 10%.

546   Undescended testes.
Note: Hypoplastic, empty scrotum.
Other features: Risk of sterility and neoplasia if not corrected.
Inheritance: About a 5% recurrence risk for sibs and offspring.

16 Endocrine disorders

547 Growth hormone deficiency.
Note: Severe short stature, mid-facial hypoplasia, prominent foreheads in offspring of consanguineous parents (these children are cousins from a highly inbred pedigree).

Inheritance: Most commonly autosomal recessive. X-linked recessive and autosomal dominant families have been described.

548

549

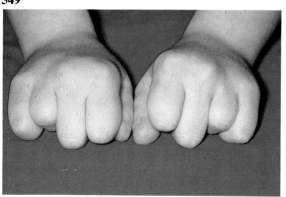

550

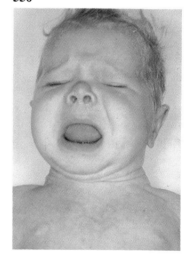

548 and 549 Pseudohypoparathyroidism.
Note: Round facies, short neck and short 4th and 5th metacarpals.

Other features: Short stature, mental retardation, calcification of basal ganglia, tetany, convulsions and hypocalcaemia.

Inheritance: Probably X-linked dominant, although occasionally male to male transmission has been reported.

550 Congenital hypothyroidism.
Note: Coarse facial features, large tongue and coarse hair.

Other features: Hoarse cry, large fontanelle, umbilical hernia and mental retardation if untreated.

Inheritance: Most cases sporadic but distinct genetic types caused by biochemical defects exist and are autosomal recessive.

551

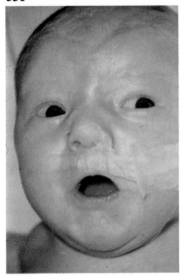

551　Di George syndrome.
Note: Hypertelorism, micrognathia.

Other features: Antimongoloid slant to eyes, notched pinnae, thymic aplasia leading to T-cell deficiency, hypoparathyroidism leading to neonatal hypocalcaemia and congenital heart defects (mostly conotruncal). This syndrome is caused by a defect in the development of the 3rd and 4th branchial arches.

Inheritance: Most cases are sporadic but rare autosomal recessive families have been reported.

552

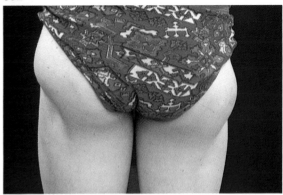

552　Diabetes.
Note: Lipo-atrophy secondary to insulin injections.

Other features: Diabetes is not a homogeneous entity and has been divided into several subtypes. For juvenile onset diabetes the recurrence risk for sibs is between 5–10% with similar risks for offspring. If an individual shares identical HLA types with an affected family member these risks may be higher. In maturity onset diabetes the recurrence risk to sibs is between 10–25%.

553

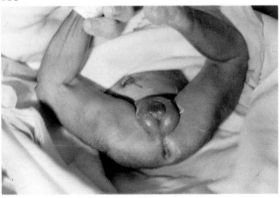

554

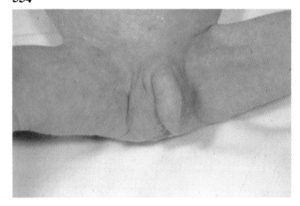

553 and 554　Congenital adrenal hyperplasia.
Note: Enlarged genitalia in a male. Enlarged clitoris and pigmentation in a female.

Other features and inheritance: Clinical features, including salt retention or loss depend on the specific enzyme deficiency involved. The most common type is 21-hydroxylase deficiency which is linked to the HLA region. All types are autosomal recessive. Prenatal diagnosis should be considered.

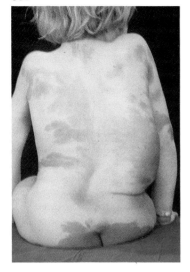

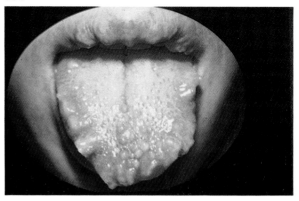

556 Multiple mucosal neuromas.
Note: Multiple neuromas of the tongue and lips. The lips are also prominent.

Other features: Medullated nerve fibres of cornea, marfanoid habitus. Medullary carcinoma of the thyroid, phaeochromocytoma.

Inheritance: Autosomal dominant.

555 McCune–Albright syndrome.
Note: Large patches of pigmentation with irregular borders ('coast of Maine appearance'), scoliosis.

Other features: Polyostotic fibrous dysplasia, pathological fractures and precocious puberty.

Inheritance: Most cases have been sporadic.

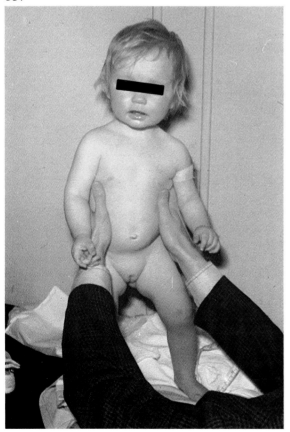

557 Testicular feminisation.
Note: Normal female habitus, scar in right groin where a testis was found in the inguinal canal.

Other features: 46, XY (male) karyotype, shortened vagina and absent uterus and end-organ unresponsiveness to testosterone.

Inheritance: X-linked recessive.

17 Gastro-intestinal disorders

558

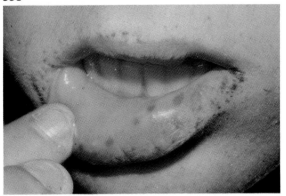

558 Peutz–Jeghers syndrome.
Note: Macular pigmentation of peroral skin and buccal mucosa.
   Other features: Polyps of small bowel, intersusseption.
   Inheritance: Mostly sporadic, some dominant families reported.

559

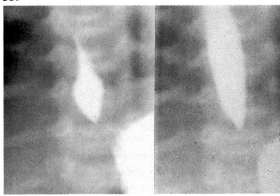

559 Oesophageal atresia.
Note: Blind ending oesophagus on X-ray barium studies.
   Other features: Tracheo-oesophageal fistulae.
   Inheritance: Usually sporadic, recurrence risk 1–2%.

560

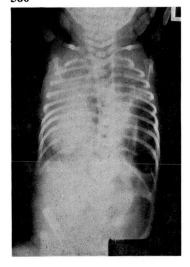

560 Diaphragmatic hernia.
Note: Intestinal contents in the left side of chest with the mediastinum shifted to the right.
   Other features: Can be associated with other congenital anomalies, such as exomphalos and neural tube defects.
   Inheritance: Recurrence risk less than 1%.

561

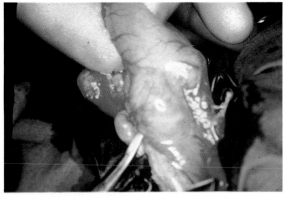

561 Pyloric stenosis.
Note: Pyloric tumour.
   Other features: There is a 4:1 male to female sex ratio.
   Inheritance: See **Table 19**.

562

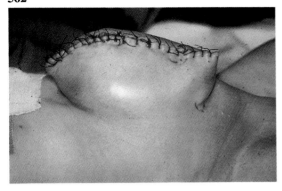

562 Exomphalos.
Note: Repaired sac which contained abdominal contents sited at the umbilical ring.

Other features: Can be part of syndromes, e.g. Beckwith–Wiedemann, trisomy 18. Prenatal diagnosis possible by raised amniotic fluid AFP and ultrasound.

Inheritance: Mostly sporadic, recurrence risk 1%.

Table 19   Pyloric stenosis, risk to sibs

Index case	Male sib	Female sib
Male	3.8%	2.7%
Female	9.2%	3.8%

Table 20   Hirschsprung disease – recurrence risk for sibs

Index case Short segment	Male sib	Female sib
Male	4.7	0.6
Female	8.1	2.9
Long segment		
Male	16.1	11.1
Female	18.2	9.1

From Harper (1981).

563

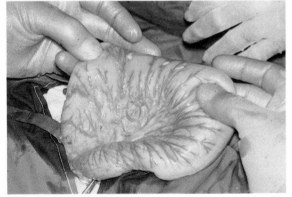

563   Hirschsprung disease.
Note: Dilated colon with narrowed aganglionic segment distally.

Inheritance: Possibly multifactorial. There is a 3:1 male: female sex ratio. Cases can be divided into long and short segment types. The long segment type is defined by the aganglionic segment extending past the sigmoid colon. For recurrence risk see **Table 20.**

564

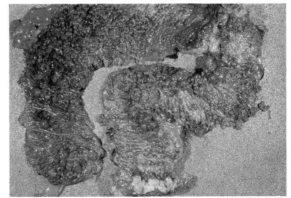

564   Polyposis coli.
Note: Multiple polyps in the colon.

Other features: Carcinomatous change. Intestinal polyps can be part of many syndromes (see **Table 21**).

Inheritance: Autosomal dominant. Individuals at risk should be screened repeatedly for manifestations of the disease. Prophylactic colectomy should be considered in affected individuals.

Table 21   Syndromes featuring intestinal polyps

Syndrome	Features	Cause
Gardner	Multiple osteomas of facial bones, epidermoid cysts and dermoids or fibromas of skin	A.D.
Turcot	Gliomas or medulloblastomas	A.R.
Peutz–Jeghers	Oral pigmentation, polyps of small intestine	Sporadic
Cronkheit–Canada syndrome	Oedema, malabsorbtion, alopecia and nail dystrophy with skin pigmentation	Sporadic

Table 22   Syndromes featuring anal atresia

Syndrome	Features	Cause
Vater	<u>V</u>ertebral <u>a</u>nomalies, <u>T</u>-<u>E</u> fistula, <u>r</u>enal and radial defects	Sporadic
G-syndrome (Opitz–Frias)	Hypertelorism, laryngeal cleft and hypospadias	? X-linked recessive
Kaufman–McKusick	Hydrometrocolpos, polydactyly and congenital heart defect	Autosomal recessive
Johanson–Blizzard	Hypoplastic alae nasae, scalp defects, malabsorbtion and hypothyroidism	Autosomal recessive
Asymmetric crying facies	Congenital heart disease	Mostly sporadic
Cat-eye syndrome	Coloboma, congenital heart defect and mental retardation	Partial trisomy 22

565

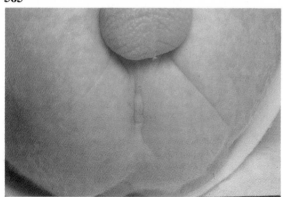

565   Anal atresia.
Note: Absence of anal opening.
Other features: Can be part of many syndromes (see **Table 22**).

566

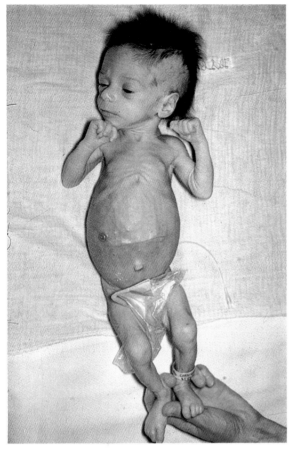

567

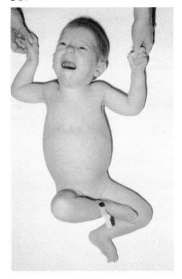

567 Coeliac disease.
Note: Abdominal distension, pale irritable child.
  Other features: Gluten enteropathy.
  Inheritance: Association with HLA A_1-B_8-Dw3 haplotype. Sibs and offspring have a 10% risk of developing the condition.

566 Cystic fibrosis.
Note: Emaciated appearance in a neonate, distended abdomen.
  Other features: Meconium ileus, steatorrhoea, chronic pulmonary disease, salt loss and dehydration.
  Inheritance: Autosomal recessive.

568

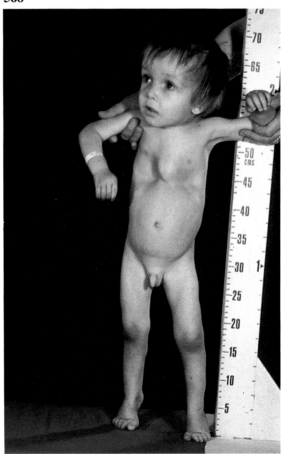

568 Shwachman syndrome.
Note: Wasted, anaemic child and gross pectus carinatum.
  Other features: Pancreatic insufficiency, metaphyseal dysplasia.
  Inheritance: Autosomal recessive.

18 Cardiovascular disorders

Table 23   Recurrence risks after an isolated defect

Congenital heart defect	Population incidence 10,000	Recurrence % Sibs	Recurrence % Offspring
Ventricular septal defect	21	3	1.9–3
Persistent ductus arteriosus	7	2.3	2.5
Pulmonary stenosis	5.5	2.2	3
Atrial septal defect	5	1.4	2.3
Transposition of great vessels	4	1.7	?
Aortic stenosis	3.5	3.1	3.9
Coarctation of aorta	3.5	1.5	2.7
Tetralogy of Fallot	3	2	4
AV defect (other than with trisomy 21)	4	1.7–8 (?3)	5–10
Truncus arteriosus	1.5	1	
Ebstein's anomaly	1.2	1	
Hypoplastic left heart	2	1	
Tricuspid atresia	1.5	1	
Primary endocardial fibroelastosis	1.7	5	
Pulmonary atresia	2	1	

569

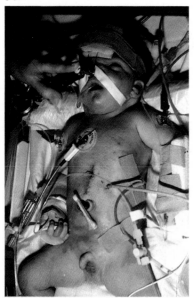

569   Isolated congenital heart defect.
See **Table 23**.

570

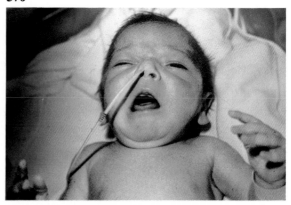

570   Asymmetric crying face.
Note: Diminished movement of mouth on left when baby is crying.

Other features: Congenital heart disease (most commonly VSD), renal and vertebral anomalies and anal atresia.

Inheritance: Mostly sporadic.

571

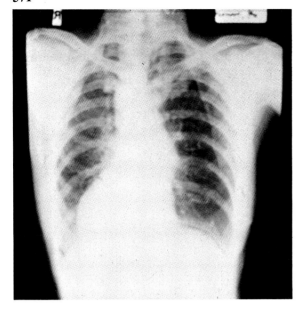

571 Kartagener syndrome.

Note: Situs inversus.

Other features: Bronchiectasis, sinusitis and sterility in males. Individuals with this disorder have been found to manifest abnormalities of cilia and sperm motility.

Inheritance: Possibly autosomal recessive, however the recurrence risk in sibships is about 10–15%.

572

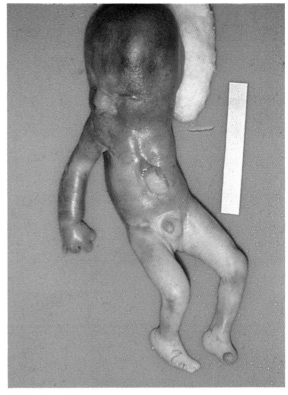

572 Acardia.

Note: Malformed fetus with absence of facial features.

Other features: Absence of heart.

Inheritance: Sporadic. These infants are usually one of monozygotic twins with a common fetal circulation.

573

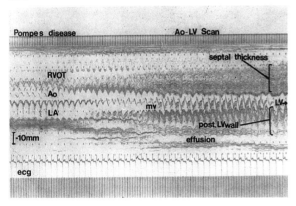

573 Pompe disease.

Note: Hypertrophy of cardiac wall and septum.

Other features: Abnormal glycogen storage, hypotonia with hepato-splenomegaly. Deficiency of acid maltase.

Inheritance: Autosomal recessive.

574

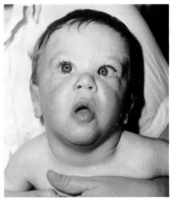

574–581 Williams syndrome.

Features: Mild to moderate mental retardation, short stature, 'Elfin' facies consisting of medial eyebrow flare, stellate pattern to iris, peri-orbital fullness, drooping cheeks, long philtrum and

575

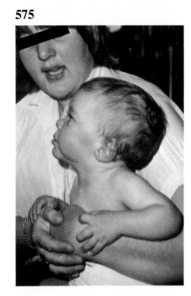

576

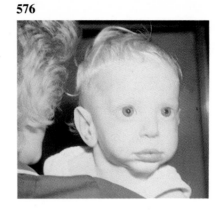

577

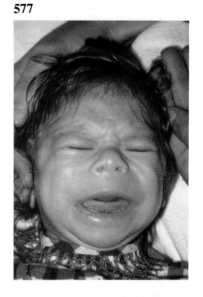

578

579

580

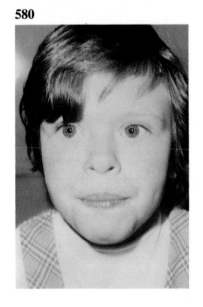

581

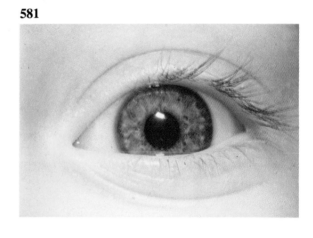

thickish lips with a tendency for the mouth to hang open. The syndrome is associated with neonatal hypercalcaemia and heart defects, especially supravalvular aortic stenosis.

Inheritance: Usually sporadic although dominant pedigrees have been described. The recurrence risk for sibs of an isolated case is about 3%.

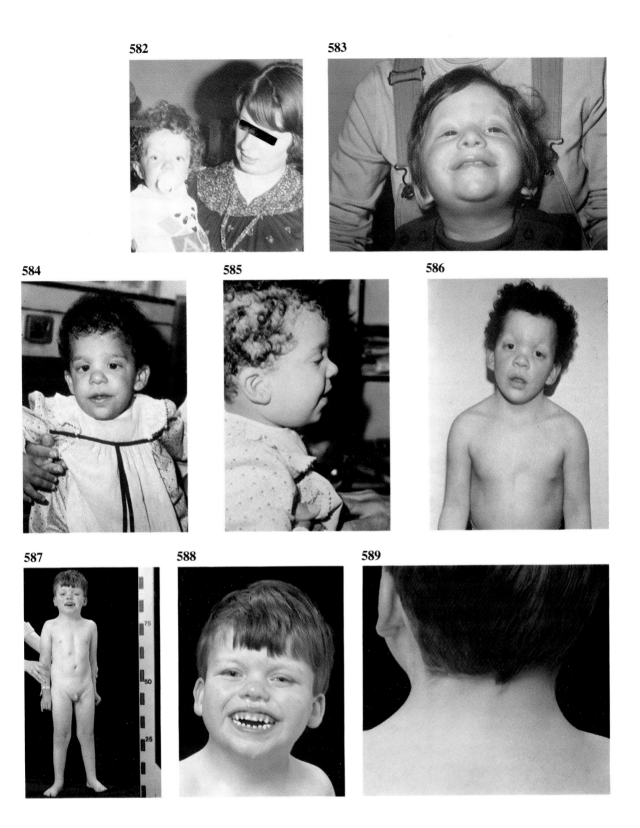

582–589   Noonan syndrome.

Note: Antemongoloid slant to eyes, ptosis, mild hypertelorism, epicanthus, low-set, posteriorly rotated ears, short webbed neck with trident hairline in some and pectus excavatum.

Other features: Short stature, mild mental retardation in some, congenital heart defect, especially pulmonary stenosis, ASD, asymmetrical septal hypertrophy.

Inheritance: Probably autosomal dominant.

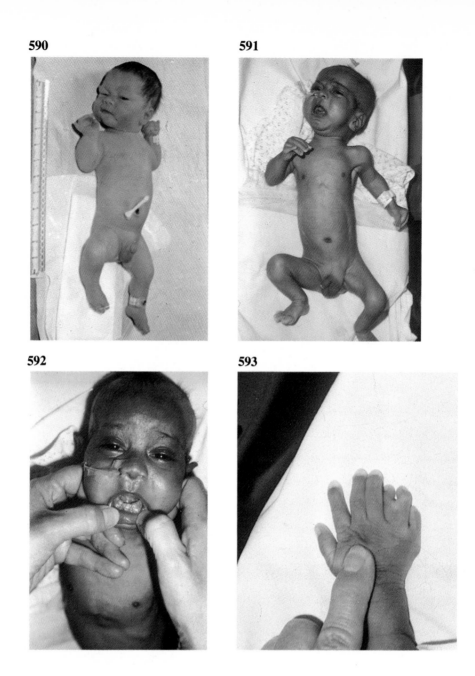

590–593 Ellis–Van Creveld syndrome.

Note:

(a) Short limbs, small chest making trunk appear long and thin.

(b) Multiple oral frenulae bound to defects of alveolar ridge, neonatal teeth.

(c) Postaxial polydactyly, hypoplastic nails.

Other features: Congenital heart defects (mostly ASD), partial anodontia, characteristic 'trident' appearance of acetabular roof, hypoplasia of upper lateral tibiae, fusion of capitate and hamate and epispadias.

Inheritance: Autosomal recessive.

594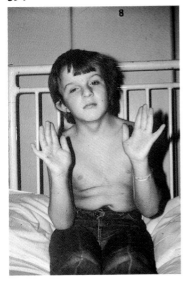

595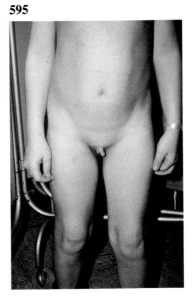

594 and 595   Charge syndrome.
Note:
(a) Deafness (hearing aid), congenital heart disease (repair), 5th finger clinodactyly.
(b) Hypoplastic genitalia.
   Other features: <u>C</u>oloboma, <u>h</u>eart defect, <u>a</u>tresia (choanal), <u>r</u>etardation, <u>g</u>enital anomalies and <u>e</u>ar (deafness).
   Inheritance: Mostly sporadic.

596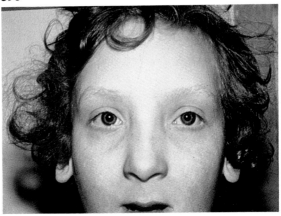

596   Alagille syndrome (arteriohepatic dysplasia).
Note: Prominent forehead, deep-set eyes with mongoloid slant and straight nose with prominent nasal bridge.
   Other features: Congenital heart disease, particularly peripheral, pulmonary stenosis, neonatal hepatitis, obstructive jaundice, mild hepatic dysfunction and mild mental retardation.
   Inheritance: Autosomal dominant with very variable expression.

19 Blood disorders

597

598

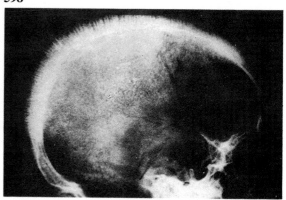

597 and 598   Sickle cell anaemia.
Note: Hypochromic anaemia with sickle and target cells.

X-ray: Note – 'hair on end' appearance of the bones of the cranial vault.

Inheritance: Autosomal recessive. Heterozygotes are very common in some parts of Africa and in peoples of African origin. Heterozygote detection is simple and should be carried out when couples of African origin marry. Antenatal diagnosis is possible both by fetal blood sampling and by DNA analysis of amniotic fluid cells. The disease is caused by a single amino-acid substitution in the ß-globin chain.

599   α -Thalassaemia.
Note: Hydrops fetalis in the severe lethal form.

Other features and inheritance: There are two ∝-globin loci. Clinical features depend on the number of ∝-globin genes absent (see **Table 24**). Antenatal diagnosis possible.

599

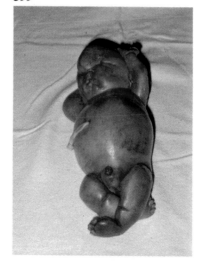

Table 24 Types of α-thalassaemia

Condition	Number of Hb α genes active	Clinical features	Main haemoglobins
Normal	4	—	A, A_2
'Silent carrier'	3	—	A, A_2
∝-Thal trait	2	Mild anaemia	A, A_2
Hb H disease	1	Severe anaemia	$H(\beta_4)\ A$
Hydrops fetalis	0	Hydrops, fetal demise	Barts (γ_4)

601

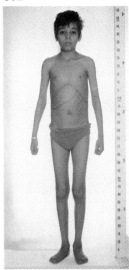

600

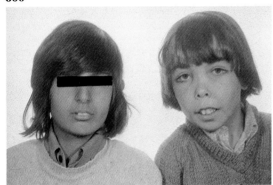

600 and 601 ß-Thalassaemia.
Note: Enlarged liver and spleen due to extra-medullary haematopoiesis. The face on the right shows broadening of nasal bridge and alveolar ridge.

Other features: Anaemia, haemosiderosis and demise in the second decade.

Inheritance: Autosomal recessive, there are different molecular forms. Antenatal diagnosis possible by fetal blood sampling and DNA analysis of amniotic fluid cells in some cases.

602

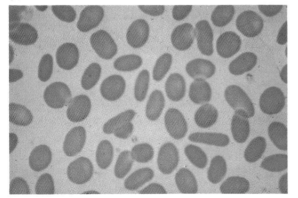

603

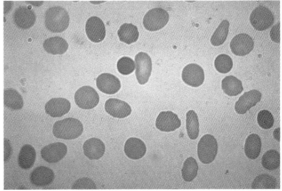

602 Elliptocytosis.
Note: Large, elliptic red cells.
Other features: Haemolytic anaemia.
Inheritance: Autosomal dominant. There are two distinct forms, one of which is linked to the rhesus blood group locus.

603 Spherocytosis.
Note: Spherical red cells.
Other features: Haemolytic anaemia, spleno-megaly.
Inheritance: Autosomal dominant.

604 Haemophilia A.

Note: Swollen knee due to haemarthrosis.

Other features: Severe bruising tendencies. Progressive joint deformity if untreated. Factor VIII deficiency.

Inheritance: X-linked recessive. Carrier detection tests possible but are not 100% reliable. Antenatal diagnosis possible by fetal blood sampling.

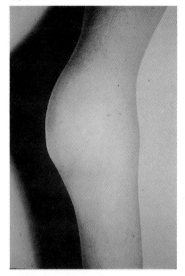

604

605 Thrombocytopenia – absent radius (TAR) syndrome.

Note: Radial aplasia, purpura.

Other features: Congenital heart defect, defects of lower limbs.

Inheritance: Autosomal recessive.

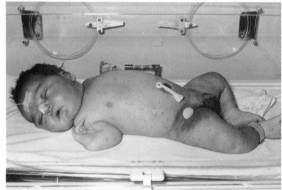

605

606 Wiskott–Aldrich syndrome.

Note: Eczema over shoulder.

Other features: Thrombocytopenia, immunodeficiency and increased risk of malignancies.

Inheritance: X-linked recessive.

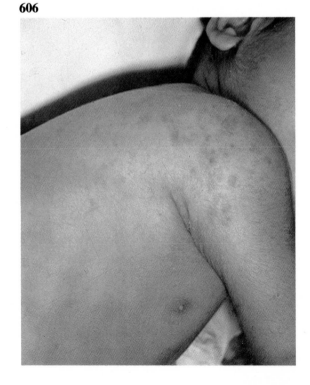

606

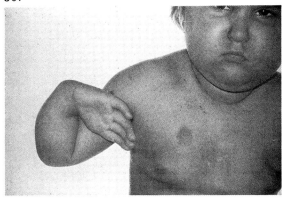

607 Fanconi anaemia.

Note: Blotchy brownish pigmentation of the skin, radial aplasia.

Other features: Short stature, pancytopenia and mental retardation in some. Enhanced chromosome breakage with a predisposition to develop leukaemia.

Inheritance: Autosomal recessive.

References

Useful books

Baraitser, M. (1982). *The Genetics of Neuro-logical disorders*. Oxford University Press.

Beighton, P. (1978). *Inherited Disorders of the Skeleton*. Churchill Livingstone.

Der Kaloustian, V.M. and Kurban, A.K. (1979). *Genetic Diseases of Skin*. Springer–Verlag.

Goodman, R.M. and Gorlin, R.J. (1977). *Atlas of the Face in Genetic Disorders*. 2nd ed. C.V. Mosby Company.

Gorlin, R.J., Pindborg, J.J. and Cohen, M.M. Jr. (1976). *Syndromes of the Head and Neck*. 2nd ed. McGraw–Hill.

Konigsmark, B.W. and Gorlin, R.J. (1976). *Genetic and Metabolic Deafness*. W.B. Saunders Company.

McKusick, V.A. (1972). *Heritable Disorders of Connective Tissue*. 4th ed. C.V. Mosby Company.

McKusick, V.A. (1978). *Mendelian Inheritance in Man*. 5th ed. John Hopkins.

Smith, D. (1982). *Recognizable Patterns of Human Malformation*. 3rd ed. W.B. Saunders Company.

Spranger, J.W., Langer, L.O. and Wiedemann, H.-R. (1974). *Bone Dysplasias*. W.B. Saunders Company.

Temtamy, S.A. and McKusick, V.A. (1978). *The Genetics of Hand Malformations*. Alan Liss.

General books on genetics

Emery, A.E.H. (1979). *Elements of Medical Genetics*. 5th ed. Churchill Livingstone.

Fraser Roberts, J.A. and Pembrey, M.E. (1978). *An Introduction to Medical Genetics*. Oxford University Press.

Harper, P.S. (1981). *Practical Genetic Counselling*. John Wrights.

Murphy, E.A. and Chase, G.A. (1975). *Principles of Genetic Counselling*. Year Book Medical Publishers Inc.

Vogel, F. and Motulsky, A.G. (1979). *Human Genetics: Problems and Approaches*. Springer–Verlag.

Index

All numbers refer to page numbers.